Kimberley monsoon rainforests
ISLANDS IN A SEA OF SAVANNA

BY KEVIN F KENNEALLY AM

UWA PUBLISHING

"A foreigner can photograph the exteriors of a nation, but ... no foreigner can report its interior – its soul, its life, its speech, its thought ... [A] knowledge of these things is acquirable in only one way ... years and years of unconscious absorption ... One learns peoples through the heart, not the eyes or the intellect."

MARK TWAIN

Rising above St George Basin in the Prince Regent National Park are two prominent flat-topped sandstone mesas underlain with basalt (Carson Volcanics). The Traditional Owners of this region, the Ngarinyin, Worora and Wunambal peoples, call them Ngayaanggananya (also known as Mount Trafalgar, in the foreground of this image; 391 metres high) and Ngorlawuroo (Mount Waterloo, emerging behind Mount Trafalgar in this image; 344 metres) (see page 185). *Image: Mike Donaldson*

Torres Strait Islander and Aboriginal peoples are advised that this book contains images and stories of people who have passed away. Permission has been given by the relevant Traditional Owners for the use of this material. Any person to whom this may cause distress or offence should exercise care in reading this book.

I acknowledge and pay my respects to all Indigenous and Traditional peoples of the Kimberley, especially those who have been extremely generous over many decades in conducting me through Country and sharing their vast knowledge, wisdom, culture and beliefs – all of which has given me a far greater appreciation of this ancient landscape that is their home.

Species names

Each species is described using its scientific name (in Latin and italics) and, where possible, its common name. Indigenous names are denoted by italics.
In cases where a species has more than one common or Indigenous name, I have listed

Images

All images are by the author unless otherwise indicated. Copyright of images is retained by the owner/

ABOVE

Wanjina. *Image: Russell Ord for Wunambal Gaambera Aboriginal Corporation*

OPPOSITE

Catherine Goonack. *Image: Russell Ord for Wunambal Gaambera Aboriginal Corporation*

Foreword

I am delighted to introduce Kevin Kenneally's new book on rainforests of the Kimberley. My family and I live at Kandiwal Community at Ngauwudu (Mitchell Plateau) where we live close to wulo (rainforest patches). Some wulo places have important cultural stories and they also hold culturally important animals and plants. I am also chairperson of Wunambal Gaambera Aboriginal Corporation who are responsible for managing the Uunguu Indigenous Protected Area which includes the majority of the Kimberley's rainforest patches including the largest patches at Bariaba (Bougainville Peninsula). Our Uunguu Rangers manage fire, weeds, feral animals and other threats to wulo and our corporation is also working with Traditional Owner families to address issues related to development pressure and to apply conservation zoning to further protect wulo in our country. Through our Uunguu Visitor Pass we also welcome everyone to visit and enjoy our wulo places both coastal and inland.

Kevin has a long history of working on our country and working with our old people. His botanical knowledge and experience is unsurpassed. He first encountered rainforest patches on our country in the 1970s when he joined the WA Herbarium and went on to take part in a number of significant biological surveys in the Prince Regent River area and Mitchell Plateau followed by a dedicated Rainforest survey from 1987 to 1989. These surveys helped raise scientific awareness of our wulo and their conservation significance. Importantly for us, Kevin and his colleagues often collaborated with our elders on these biological surveys. My uncle, the late Mr Geoffrey Mangolamara, contributed a chapter on Wunambal knowledge to the Kimberley Rainforests book published in 1991 following the survey. Kevin has kindly shared some photos of elders he has worked with over the years from his personal collection.

This new book is a valuable resource for everyone who lives, works or travels in the Kimberley, pulling together all of the available knowledge of rainforest into one place. It shows us how our local species are related to rainforest plants across Northern Australia and beyond our shores. Importantly, it also recognises the cultural importance of rainforest to Traditional Owners across the Kimberley and the important role we play in looking after country. I am sure you will treasure this book as much as I do.

Goonack.

CATHERINE GOONACK
CHAIR, WUNAMBAL GAAMBERA
ABORIGINAL CORPORATION

The Cockburn Range has remnant patches of rainforest on scree slopes. *Image: Philip Schubert/Shutterstock*

"To Aboriginal people it is not about dates, it started from the Lalai (Dreaming or Dreamtime) when there was no beginning in time, it was when the world started."

JANET OOBBAGOOMA, QUOTED FROM *WE ARE COMING TO SEE YOU* (2018)

Bunjani (Little Mertens Falls), Mitchell River National Park. *Image: Lochman Transparencies*

Contents

The Banded Honeyeater (*Certhionyx pectoralis*) is common in closed-canopy riverine vegetation dominated by paperbarks, such as the Silver-leaved Paperbark (*Melaleuca argentea*). Image: Janelle Lugge/Shutterstock

INTRODUCTION

The plant communities of the Kimberley, including Australia's tropical rainforests, are the most biodiverse ecosystems of Australia and, in many ways, the least understood.

Tropical rainforests are among the most complex and species-rich ecosystems ever to have existed. Of all Earth's ecosystems, tropical rainforests exist at the extremes of temperature, rainfall, biodiversity and structural complexity. Once covering 14% of the Earth's surface, they now make up only 6%. Since 1947, the total area of tropical rainforests has been reduced probably by more than half, to about 6.2 to 7.8 million square kilometres.[1]

Tropical rainforests are characterised by enormous numbers of species and life forms. Such luxuriance can usually only be sustained by high rainfall and a tropical climate.[2] The evolution of rainforest hyper-diversity is not well understood.[3] Although the intense diversification of species in tropical rainforest is often suggested to have evolved over a long time, there is recent evidence of rainforests expanding rapidly from ancestral species into a multitude of new forms. Important climatic predictors of plant diversity at the global scale are related to energy and water availability, which suggests that the highest diversity will be found in the wet tropics, where the largest number of plant species can co-exist and therefore evolve. In the tropics, the amount and timing of precipitation strongly influence the distribution of both ecosystems and plant species richness.[4]

Across northern Australia, tropical rainforests and the animals living with them face many environmental threats, including intense wildfires, feral animals, introduced ants, weed invasion, disturbance to rainforest aquifers, and climate change.[5] Given the multiplicity of these threats, management and conservation of these biodiverse, generally remote and naturally fragmented patches of rainforest are not easy tasks, particularly in the Kimberley. Just recognising the value of rainforests in the landscape is not sufficient. They need legislative protection to ensure their conservation and survival.

Indian Prickly Ash (*Zanthoxylum rhetsa*). *Image: Tim Willing*

A sound understanding of ecosystem structures and processes, as well as systematics, is a prerequisite to good conservation policy. This is especially true in poorly studied areas where living creatures are exterminated before they can be discovered and described.

When I published the first checklist of Kimberley vascular plants in 1989, only 1860 species of angiosperms (flowering plants), gymnosperms (non-flowering plants) and pteridophytes (ferns and fern allies) had been recorded.[6] The Western Australian Herbarium's Florabase now lists around 3500 species of angiosperms for the Kimberley, and several hundred more species have been discovered but not yet formally described.[7] Many of these newly discovered species have been found only in the Kimberley, can be locally restricted, and may be rare or threatened. The first step to ensuring their conservation is to describe them formally. At the same time, we urgently need further research on the systematics, ecology and conservation of all Kimberley flora.

Many plant species in the Kimberley rainforests have been traditionally used by Aboriginal people in the region and by other cultures in South-East Asia as sources of food as well as pharmaceutical and medicinal compounds. The full potential of these plants has yet to be investigated.

Graham Donation and Martina Dixon in 1988, up a *Gubinge* tree in Broome.

In addition, several species of fruit-eating birds and mammals move seeds between tropical rainforest patches and require many patches to maintain their populations. Consequently, every patch has value, and we cannot afford to lose these 'jewels in the crown' scattered throughout the savanna woodlands of the Kimberley. Using the precautionary principle, we need to balance rainforest conservation with present and emerging uses in the Kimberley.[8]

This book reviews what we do and don't know about Kimberley rainforests and why we need to acknowledge their value, not only to the biodiversity of northern Australia but also in terms of their traditional use and cultural significance to Aboriginal peoples, as well as their relationship to the monsoon rainforests of South-East Asia. During my many years of studying the flora and vegetation of the Kimberley (including rainforests), I have benefitted greatly from collaboration with Indigenous Australians. I have sat with and listened to Aboriginal people as they shared their extensive knowledge and culture, and as they guided me through Country and allowed me to document and photograph this journey.

Boca de Valeria, a village on the edge of the Amazon rainforest, Brazil.

Logging in the Amazon rainforest in the Tapajós National Forest, Brazil.

CHAPTER 1
Tropical rainforests

Rainforest canopies are generally so dense that they let very little sunlight reach the ground. As a result, a humid and relatively stable micro-climate develops in the understorey that supports and protects a myriad of lifeforms. Falling trees disrupt the integrity of the forest canopy, creating various regeneration niches.

Tropical rainforests store large amounts of carbon, but agreement is lacking on their net contribution to the terrestrial carbon balance. A study of tropical forests in South America, Africa and Asia (between 23.45°N and 23.45°S, excluding Australia) showed that tropical rainforests - which until recently had a key role in absorbing greenhouse gases - are now, as a result of large-scale deforestation since the 1960s, releasing 425 billion kilograms of carbon annually.[1] This is more than the annual carbon generated by all the vehicles in the USA.[1-3] Since the 1960s, disturbance has reduced the biomass in these rainforests by up to 75%. As the lead author of the study explained: "As always, trees are removing carbon from the atmosphere, but the volume of the forest is no longer enough to compensate for the losses. The region is not a sink anymore."[4] The researchers concluded that the priority is to protect pristine forests with high carbon density. The most effective way of doing this, they said, was to support land rights for Indigenous people.

The world's largest rainforest is the Amazon, which spreads across nine countries, covers nearly 40% of South America, and accounts for just over half the primary forests in the tropics. Over one-third of all plant species in the world live in the Amazon rainforest, making it the most biodiverse tropical forest in the world. An area greater than Australia's entire rainforests is cleared annually in the Amazon basin.[5] Since 1978, about one million square kilometres of Amazon rainforest have been destroyed across Brazil, Peru, Colombia, Bolivia, Venezuela, Suriname, Guyana and French Guiana.[6]

TROPICAL RAINFORESTS IN AUSTRALIA

Although rainforests are generally associated with the evergreen jungles of South America, Africa and South-East Asia, Australia has about 36,000 square kilometres of tropical rainforest – 3% of the country's total forest area. Since European settlement in 1788, the country has lost 27% of its rainforest (Fig. 1).

The tropical rainforests found across northern Australia are the same as monsoon forest as defined by German botanist Andreas Franz Schimper in 1903: "as more or less leafless during the dry season, especially towards its termination, is tropophilous in character (adapted to a climate characterised by marked environmental changes), usually less lofty than rainforest, rich in woody lianas, rich in herbaceous, but poor in woody epiphytes".[7] In northern Australia, tropical rainforest includes semi-deciduous monsoon

FIGURE 1   Map showing the distribution and size of monsoon rainforest patches across northern Australia. *Image: Jeremy Russell-Smith*

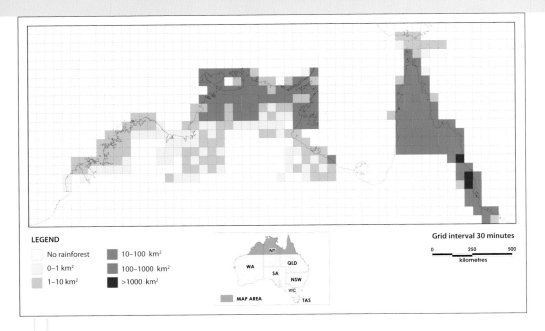

LEGEND

No rainforest	10–100 km²
0–1 km²	100–1000 km²
1–10 km²	>1000 km²

MAP AREA

Grid interval 30 minutes

0   250   500
kilometres

vine thickets, which are distinguished from leafy 'evergreen' (non-deciduous) forests by becoming more or less leafless (deciduous or semi-deciduous) during the dry season and 'raingreen' during the wet season.

Seasonally dry tropical rainforests, such as those in the Kimberley, also occur in the Americas, Africa, India and elsewhere in Australia. They sustain large biotic populations, determine regional climate, are sites of biological and cultural conservation, and have significant economic value.[8]

Australia's monsoon rainforests are part of a great corridor of monsoon forests that extend through the areas of Australasia that have strong seasonally wet and dry climates. These rainforests are thought to be part of a formerly more widespread palaeotropical flora that has evolved through a series of marked paleoclimates, possibly through the inland adaptation of plants that grow along shorelines.[5]

Monsoon rainforest in Australia now occurs as small, seasonally sparse, raingreen patches that are confined to gullies and scree slopes or located behind coastal sand dunes. Many monsoon forests occur on sites where soil moisture is higher (such as alongside springs, seeps and creek lines). In some cases, they are protected from fire by the rock or boulder fields on which they grow.[9,10] Their flora is similar to that of the evergreen rainforest, although the distinctive structural features of the evergreen (or humid) rainforest are sparse or absent, and annual herbs may be present.[11]

The vegetation in Kimberley monsoon rainforest patches can be classified structurally and floristically into three types based on the availability of moisture: complex mesophyll monsoon forest, semi-evergreen mesophyll monsoon forest, and deciduous monsoon vine thicket.[12]

Monsoon rainforests are one of only three closed-canopy vegetation communities found in the wet-dry tropics of the Kimberley; the others are mangroves and riparian, or gallery, forests that fringe the banks of major creeks, rivers and sandstone gorges. These communities are important as biological corridors and are essential habitat for some species.

Scattered throughout the vast areas of savanna in the north-west Kimberley, small patches of monsoon rainforest have similar flora to humid rainforests of South-East Asia and north-east Queensland.[13] Large tracts of humid evergreen rainforest are now restricted to the north-east coast of Australia and are separated from the monsoon forests of Western Australia and the Northern Territory by the Gulf of Carpentaria and the arid grasslands of the Barkly Tablelands.[14]

Australian rainforests have very few species in common with the adjacent sclerophyll vegetation. They are distinguished from other closed-canopy forests by the prominence of trees, shrubs and vines bearing fleshy fruits, epiphytes, lianas (woody vines), aerial roots and buttress tree trunks.[15]

The high diversity of plant species in the Kimberley monsoon rainforests is characterised by large numbers of families and genera that each contain only one or a few species.[11] This diversity is a contributing factor in the high level of complexity found in Kimberley monsoon rainforests. Because many of these plant families and genera do not occur outside these rainforest patches, and they currently have no protection under state or federal legislation, their conservation is crucial.

These raingreen monsoon rainforests are characterised as much by the relatively high proportion of mobile, wide-ranging plant and animal species as by their largely deciduous canopy. Because trees lose moisture through their leaves, the shedding of leaves during the dry season allows trees to conserve water. Such shedding also opens up the canopy, enabling sunlight to reach the ground and facilitate the growth of a dense understorey. A study of litter fall in monsoon rainforests and mangroves at Mitchell Plateau revealed pronounced seasonal variations in leaf fall, which was monophasic at the Walsh Point rainforest patch (occurring from April to July) and biphasic at the Lone Dingo rainforest patch (occurring from April to July and from September to November).[16] However, over the entire year, the mangrove communities adjacent to Mitchell Plateau had the highest litter fall of all sites sampled.

JUNGLE OR RAINFOREST?

The terms 'jungle', 'jungle-like thickets', 'corridor jungle' and 'miniature jungle' were used in Australia by early explorers and writers such as Ernestine Hill to describe dense, impenetrable vegetation in the Kimberley and northern Australia.[17,18] More often than not, the term was applied to the dense riparian vegetation growing along creeks and rivers, occasionally to mangroves, and rarely to monsoon rainforest patches.[19] The compound noun 'rain forest' is routinely used elsewhere in the world, but in Australia it has been contracted into one word. This contraction was proposed by research forester George Baur[20] because it denotes a discrete vegetation formation and avoids what he described as the "undue emphasis on rain as the sole determining environmental factor".[21]

The word 'jungle' conjures up images of Edgar Rice Burroughs' hero Tarzan swinging from vine to vine through dense and luxuriant vegetation surrounded by a host of exotic animals. It originates from the Hindi word jangal, from the Sanskrit jāngala, meaning wilderness or uncultivated land, with the Anglo-Indian interpretation leading to its connotation as a dense tangled thicket.[22] 'Jungle' carries connotations of untamed and uncontrollable nature and isolation from civilisation, along with emotions that evoke threat, powerlessness, disorientation and immobilisation[23] – something that those of us who have explored monsoon rainforest will attest to!

Mature evergreen rainforest is usually not difficult to penetrate.[24] However, where light penetrates the canopy, the vegetation is sufficiently dense and impenetrable to hinder movement by humans, requiring travellers to cut their way through. In modern usage, the word 'rainforest' has superseded jungle as a more specific and scientific term when referring to tropical forests, although the term jungle is still used. In scientific literature, the evergreen forest is referred to as tropical rainforest, a term introduced in 1903.[7]

Kevin Kenneally in the Tapajós National Forest, within the Amazon rainforest.

Rainbow over Oomarri (King George Falls). *Image: Tim Willing*

CHAPTER 2:
The Kimberley

The Kimberley covers more than 423,000 square kilometres of north-west Australia. Its biogeographical landscape, which features massive waterways and river systems, has been shaped by thousands of years of geological history.

During the Last Interglacial (a period spanning 130,000-115,000 years ago), sea levels were 6-10 metres higher than their present levels. Thereafter, they fluctuated irregularly, falling to 120 metres below present-day levels during the Last Glacial Maximum, around 20,000 years ago. This exposed large areas of two continental shelves: the Sunda Shelf, which extends south from mainland South-East Asia, and the Sahul Shelf, which stretches from the northern coast of Australia to New Guinea. This enlarged continent, known as Sahul,

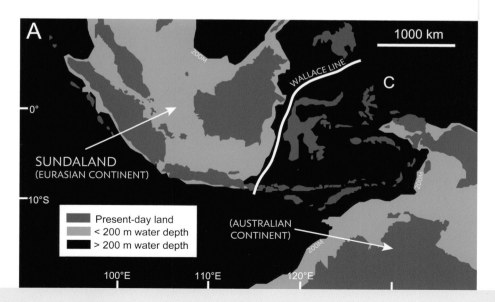

Map of Australia during the last ice age, 21,000 years ago, showing the size of the continent and its connections with South-East Asia (including Wallacea) and New Guinea. *Image: David Haig*

FIGURE 2

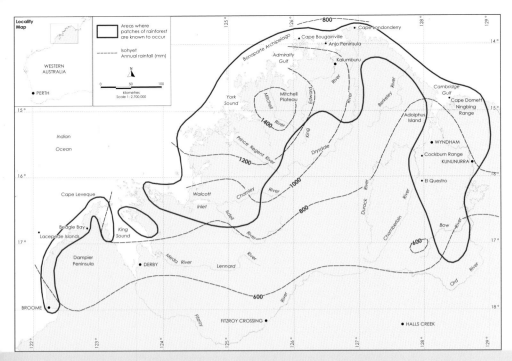

FIGURE 3   Map showing the distribution of monsoon rainforests and the average annual rainfall in the Kimberley. Adapted with permission.[5]

connected New Guinea, some small islands of modern-day Indonesia, mainland Australia and Tasmania. At its largest, Sahul was 20% larger than Australia is today, and what is now the Kimberley was less than 100 km from Timor and almost twice its current size (Fig. 2). Around 12,000 years ago, the coastline to the north-west had advanced inland 300 km over the Sahul Shelf. Many generations of Kimberley coastal Aboriginal populations experienced a continuing loss of territory over those millennia.[1]

Changes in sea level are important drivers of tropical climate change on glacial-interglacial timescales. Exposure of the Sunda and Sahul shelves caused changes in atmospheric circulation and in reflected solar energy.[2] Evidence from across much of Australia suggests that the last ice age was arid and windy and lasted approximately 6000 years.[3]

Immediately north of the Kimberley, Indonesia has vast expanses of tropical rainforests. The islands of southern Wallacea, in eastern Indonesia, are the Kimberley's nearest neighbour and a recognised global biodiversity hotspot. Deep water channels separate the islands from each other and from the Asian and Australian continental shelves. These channels represent a transition zone between the great rainforest blocks associated with the Indomalayan (Sumatra, Borneo,

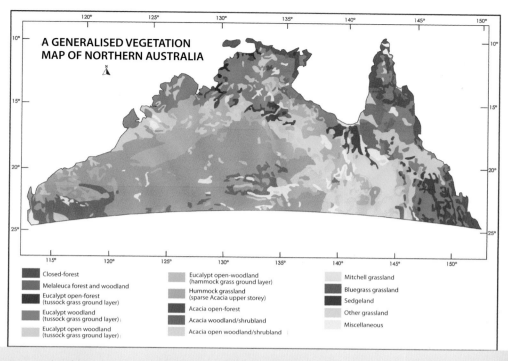

A generalised vegetation map of northern Australia. *Image: Paul Sawers*

FIGURE 4

Java) and the Australasian (New Guinea, northern Queensland) biogeographic regions. The Indo-Australian Archipelago is biogeographically complex, because it facilitates the range expansion of the two faunal assemblages occupying the continental areas on either side of it.[4]

There are more than 1500 patches of monsoon rainforest totalling 70 square kilometres scattered across 170,000 square kilometres of the north-west Kimberley (Fig. 3). They now exist effectively as islands in a sea of savanna woodlands. Kimberley monsoon rainforests are isolated, small and relatively ancient; therefore, they are evolutionarily significant ecosystems where processes and species have had the opportunity to evolve in isolation. They are embedded within a mosaic of forest or woodland savanna that is dominated by mostly flammable species of *Eucalyptus* and *Corymbia* (both of the Myrtaceae family) (Fig. 4). The average patch has an area of less than 0.04 square kilometres; only around 3% exceed 0.2 square kilometres, with the largest being 2.2 square kilometres on the Bougainville Peninsula. Mapping of more than 6000 rainforest patches on Wunambal Gaambera country in the North Kimberley showed that most of these patches are small (less than 0.01 square kilometres) and occur predominantly on nutrient-rich substrates (e.g. basalt) and fire-sheltered topographic settings.[5] Due to their small size, these rainforest patches are essentially undetectable at the resolution used by global-level assessments.[6,7] However, they are rich in species not found in the region's other vegetation communities. Of the 1500 monsoon rainforest patches in the Kimberley, fewer than 100 have been scientifically surveyed and documented.

Large groups of people – the ancestors of Indigenous Australians – first entered Sahul at least 65,000–50,000 years ago, although the exact timing is not clear.[8] The arid environment discouraged the preservation of artefacts, and rising sea levels since then have left relatively few reliable archaeological sequences. This colonisation of Australia by culturally modern humans can be considered the first true migration, as opposed to the dispersals that happened before it.[9]

The cultural landscape of the Kimberley is shaped by Aboriginal peoples, who have continuously occupied the region since their ancestors' arrival.[10-13] For at least the last few hundred years, they traded and made contact with the outside world, even travelling back and forth to Indonesia.[14] Prior to European settlement of the region, Macassan fishers from the islands of Roti (part of the East Nusa Tenggara province of eastern Indonesia) and Macassar (Ujung Padang, a port city on Indonesia's Sulawesi Island) visited Australia, including the Kimberley coast (known to Macassans as Kaju Jawi). They journeyed in large fleets of wooden sailing vessels, called praus, to harvest an edible holothurian bêche-de-mer, also known as trepang or sea cucumber (Fig. 5). Evidence of their visits to the Kimberley includes stone hearths where trepang was boiled, pottery shards and the introduced Tamarind tree (*Tamarindus indica*), whose fruits are used in Indonesian cooking (Fig. 6).

Indonesian fishermen, descendants of the Macassans, continue the tradition of harvesting trepang and clam meat in Australian waters, often illegally. These fishing boats were photographed in the Ashmore Reef Marine Park in 1977.

FIGURE 5

A member of the Landscope research expedition recording historical evidence of Indonesian fishermen visiting the Kimberley coast and processing trepang. This includes the stone cooking hearths found scattered along the coast and the Tamarind trees (*Tamarindus indica*).

FIGURE 6

Gwion are variously 'Wanjina's mate' or 'helpers' and sometimes boss of the Wanjinas and are said to travel with the Wanjina. *Image: Wunambal Gaambera Aboriginal Corporation*

CHAPTER 3
Indigenous biocultural knowledge of Kimberley rainforests

Anthropological research indicates that Aboriginal people had a particularly complex set of relationships with plants.[1] In particular, Kimberley rock art stands out globally as having an enormous body of direct and indirect depictions of plants, including grasses, trees and yams, as well as images of plant-based material culture such as digging sticks (Fig. 7).[1] Studies have described as many as 106 plant food species in Kalumburu in the North Kimberley[2,3] and 127 plant food species on Kimberley islands.[4] Most of the plant foods collected from rainforests appear to have been used for casual consumption (Figs 8, 9).

Depictions of plants are found on Aboriginal rock art galleries in the north-west Kimberley.

FIGURE 7

FIGURE 8

Broome local Graham Donation collecting the ripe fruits of *Gubinge* (*Terminalia ferdinandiana*) in 1984.

FIGURE 9

Watson (Watty) Ngerdu with a specimen of *Gubinge*. Watty was a community Elder and a leader of the Mowanjum community at Wotjulum. In 1983, he participated in the Australian and New Zealand Scientific Exploration Societies (ANZSES) expedition to Walcott Inlet and provided Aboriginal names for plants and animals.

FIGURE 10

Botanist Dr Bernie Hyland, CSIRO, discussing traditional names for rainforest plants with Wunambal Gaambera Elder Geoffrey Mangolamara during the Kimberley rainforest expedition of 1987.

FIGURE 11

Esther and Sandy Paddy (back row) with Molly Wiggan, their great-grandson Alan Bin Hitam and Audaby (Udibi) Jack, in 1984 at their campsite at Anbaraminyan near Gnamagun Well, 6 km south of Cape Leveque on the Dampier Peninsula.

Geoffrey Mangolamara with a specimen of *Lulurr* or Candelabra Wattle (*Acacia holosericea*) at a Mitchell River campsite in 1987.

FIGURE 12 (LEFT)

Donald Langi, an Elder from Mowanjum, standing next to a Boab or *Bodgurri* tree (*Adansonia gregorii*). He and Neville Morlumbum accompanied a scientific expedition along the Kimberley coast and islands in 1990 and provided Aboriginal place and plant names.

FIGURE 13 (RIGHT)

To Aboriginal people, Country is a complex living cultural landscape. It is imbued with enormous meaning and relevance to their traditional lifestyle. Aboriginal peoples' connection to Country is influenced by their belief systems, long occupation, and an extensive knowledge of natural cycles, all of which relate to both extended and more local stretches of Country.[5] Indigenous land and sea management, also referred to as 'caring for Country', includes a wide range of environmental, natural resource and cultural heritage management activities undertaken by individuals, groups and organisations across Australia for customary, community, conservation and commercial reasons (Figs 10-17).[6]

As monsoon rainforests were prime food-gathering areas, Aboriginal people would not burn them but would instead protect them by burning the surrounding grassland early in the dry season. If these protective fires were lit every year, the rainforest would never burn.[7] Some patches of rainforest were also culturally sensitive law grounds.[8,9] An example in Broome is the Lurujarri Heritage Trail,

Donald Langi and Neville Morlumbum sitting beneath sacred Wanjina rock art at Raft Point in the Kimberley in 1990.

which connects specific sites (Law Grounds) and follows the land of the traditional Song Cycle. Traditional Law was encoded in the Song Cycle and has been passed down unbroken since the Dreamtime (*Bugarregarre*). The Lurujarri trail traverses 80 km of saltwater coastal terrain and embraces patches of monsoon vine thicket.

The fact that the Bardi and Goolarabooloo languages of the Dampier Peninsula include the word *budan*, meaning 'monsoon vine thicket', confirms the importance of this plant community to Indigenous people. Bardi people also call it *mayi boordan*, meaning 'bush fruit country'. Yawuru people call it *mayingan manja balu*, meaning 'plenty of fruit trees'.[10] In the North Kimberley, it is called *wulo* in Wunambal Gaambera country.[11]

Local Aboriginal people understood that patches of monsoon vine thicket function as a network and that the occurrence of 'bush fruit' differed from patch to patch. Aboriginal concepts of seasonality are far more complex than that of the simple pattern of wet-dry seasons enshrined by Europeans.

FIGURE 15
(ABOVE)

Graham Donation and Martina Dixon of Broome in 1984, with the edible seed pods of *Magabala*, *Garlarla* or Bush Banana (*Leichhardtia viridiflora* subsp. *tropica*). This scrambling vine is common in pindan and woodland in the Kimberley but rare in monsoon rainforest.

FIGURE 16
(MIDDLE)

Sitting down with Sandy Paddy and family in 1984, talking about the bush foods they gather from a rainforest patch known as Ilan on the Dampier Peninsula.

FIGURE 17
(RIGHT)

Jack Karadada, from Kalumburu at Mitchell Plateau, in 1982, demonstrating how to harvest the edible stem or pith from the fan palm *Livistona eastonii*, known to Wunambal Gaambera people as *Yalarra* or *Dangana*.

Kevin Kenneally sheltering from torrential rain in a cave at Mitchell Plateau during Tropical Cyclone Bruno in January 1982. *Image: Bruce Maslin*

CHAPTER 4
Documenting Kimberley rainforests

Botanical observation in the Kimberley district began as early as 1688 with the visit of English explorer William Dampier, who recorded a general description of some vegetation.[1,2] The first contact with the rainforest appears to have been made by botanical collector Alan Cunningham in 1819-1822, when he accompanied the voyages of Lieutenant P.P. King charting the Kimberley coast (Fig. 18).[3,4] Many of Cunningham's specimens are derived from the monsoon forest, but he left no published record of vegetation types. In his journal, Cunningham did, however, refer to the monsoon forest. On 16 September 1820, he recorded the following from the Hunter River:

> "... steep rocky ridges of hills, furrowed with channels formed by the torrents that doubtless fall from their elevations during the season of rains in the river..... through interesting brushes which clothe the declivities with a luxuriant verdure. These thickets afforded me some variety of plants ..."[4]

And again, on 17 September:

> "Thickets clothing the broad bases of the lofty towering boundary cliffs at length became interesting. In them I observed some remarkably fine large trees of *Myristica* laden with its fruit, some of which were partially burst showing the arilla, or mace, and a tree of very regular ornamental growth, having dark green opposite

FIGURE 18

Botanist Alan Cunningham was the first European to document rainforests in the Kimberley, during the hydrographic expeditions of 1819–1822 under the command of Lieutenant P.P. King on HMC *Mermaid* and HMS *Bathurst. Image: Mitchell Library, State Library of NSW*

FIGURE 19
(FAR LEFT)

Julius Brockman at age 26, photographed in Roebourne in 1876. He recorded rainforest behind the sand dunes on the Dampier Peninsula in 1879. *Image: Joan Brockman*

FIGURE 20
(LEFT)

William Robert Easton pictured in 1921, the year he led a surveying expedition to the Kimberley. *Image: Bill Easton*

FIGURE 21
(BELOW)

Botanical collector Charles Austin Gardner, photographed on the Easton Surveying Expedition of 1921. *Image: Bill Easton*

leaves, nerved as in *Calophyllum*, was very conspicuous, without fruit. The *Cryptocarya* was likewise frequent."

The first recognisable encounter with the monsoon rainforest that has been found in the literature was recorded in November 1879 by George Julius Brockman, a prominent explorer and pastoralist in the Gascoyne and Kimberley regions (Fig. 19). He made particular note of the dense monsoon vine thickets found behind the coastal sand dunes on the north end of the Dampier Peninsula:

"We turned into the beach again at sunset... and had hard work to force our way through the jungle that skirts the sea hills, having to get out our knives to cut the tangled masses of the creeper, often as strong as rope."[5]

Sir George Grey was the first to record the Pied Imperial Pigeon (*Ducula bicolor*) in Western Australia, a bird reliant on fruiting monsoon rainforest patches. Grey collected two specimens on 17 December 1837 at Hanover Bay, near the mouth of the Prince Regent River.[6] In 1910, several rainforest bird species – including the Orange-footed Scrubfowl (*Megapodius reinwardt*) and Rose-crowned Fruit-dove (*Ptilinopus regina*) – were recorded by field collector Gerald Hill in "dense tropical scrubs" composed of "tropical trees, shrubs and creepers" at Parry Harbour in the North Kimberley.[7]

These early encounters tended to be superficial and inconclusive, and it was not until 1923 that the first general account of the vegetation of the Kimberley was given by C.A. Gardner, who explored the north-west Kimberley botanically in 1921 when travelling with the expedition of W.R. Easton (Figs 20, 21).[8,9] Even then, however, Gardner failed to observe and record monsoon vine thickets and forests, writing in his account of the "total absence of rainforests". Unaccountably, he apparently never saw the numerous discrete patches in the Mitchell Plateau area, although his party traversed the plateau from end to end. There is no mention of any monsoon vine thickets and forests elsewhere, such as in the Prince Regent River, nor does the list of species found on the expedition suggest that he had collected in them.

In a later work, Gardner emphasised the prominence of an "Indo-Melanesian element" in the Kimberley flora, which he equated with the "Palaeotropic element" of the Australian flora defined by British botanist Sir Joseph Dalton Hooker, who had visited Australia as part of an Antarctic expedition from 1839 to 1843.[10,11] However, Gardner was referring generally to tropical species found in the littoral zone, riverine forest, woodlands and savannas, rather than to the monsoon forest to which the term might more appropriately relate.

FIGURE 22

Dr John Beard, a distinguished ecologist and the foundation director of Kings Park and Botanic Garden in Perth, examining a rainforest plant specimen at the Hunter River in June 1987.

Traversing of the North Kimberley by members of the Land Research Division of CSIRO in 1954 was limited by inaccessibility of the country. As they reported: "There are no rainforests, usually considered characteristic of the tropics, and even the monsoon forests described for the Katherine-Darwin region by Christian and Stewart (1953) are absent".[12,13] However, the existence of monsoon vine thickets at Mitchell Plateau was eventually brought to light in the course of a geological survey in 1965. It was carried out jointly by the Bureau of Minerals Resources and the Geological Survey of Western Australia and reported in the explanatory notes of the Montague Sound 1:250,000 geological sheet.[14,15]

In 1965, the Amax Bauxite Corporation began to investigate bauxite deposits on Mitchell Plateau. A mining camp was established and linked to the outside world by air and by vehicle track. The provision of accommodation, large work areas, a mess and access to helicopters and vehicles made for very effective and efficient fieldwork conditions. As a result, scientists of many disciplines visited the Plateau and became aware of the monsoon rainforest. The first study was conducted in 1974 by J.S. Beard, who published a preliminary description and classification along with a map showing the more important patches.[16] Beard also drew attention to the relatively more extensive occurrence of monsoon vine thickets on the Bougainville

Peninsula, which no botanist had yet visited (Fig. 22). At about the same time, P.G. Wilson and N.G. Marchant of the Western Australian Herbarium examined a number of the islands off the north-west Kimberley coast during biological survey expeditions coordinated by the Department of Fisheries and Wildlife. They reported and collected samples of plants and animals in patches of monsoon vine thicket.[17]

Also in 1974, a biological survey of the Prince Regent River Reserve in the North Kimberley was coordinated by the Department of Fisheries and Wildlife.[18] Twelve sites were recorded in detail, but no general account of the vegetation was included. However, the expedition reported finding semi-deciduous monsoon vine thickets at three of the sites, along with closed forest of Soapwood (*Alphitonia excelsa*), Alligator Bark (*Calophyllum sil*), Billabong Tree (*Carallia brachiata*), Brown Bollywood (*Litsea glutinosa*), River Cajeput (*Melaleuca leucadendra*), Australian Nutmeg (*Myristica insipida*) and Swamp Ash (*Syzygium angophoroides*) along river and creek beds. The following year, a similar survey of the Drysdale River National Park revealed the presence of three communities of monsoon semi-deciduous vine thicket and one of riverine closed forest.[19]

Related botanical surveys were made by Roger Hnatiuk and Kevin Kenneally of the Western Australian Herbarium in 1976, 1978 and 1979, and their report on the vegetation and flora of Mitchell Plateau was included in a biological survey publication from the Western Australian Museum.[20] A list and description of 26 communities (one being the monsoon vine thickets) was provided, along with an annotated list of the flora, of which approximately 80 taxa belonged to the monsoon rainforest. In 1978 and 1979, John Beard, Kevin Clayton-Greene and Kenneally conducted a survey at the Plateau for Amax Bauxite Corporation and also visited the Bougainville Peninsula.[21] In 1982, Kenneally returned to the Plateau in January, April, October and December. In April and October, he again visited Bougainville Peninsula, when he made extensive helicopter traverses and collected botanical specimens (Figs 23, 24). These visits resulted in a private report to the company, a paper on the monsoon vine thickets of the Bougainville Peninsula and another on the fire ecology of the monsoon vine thickets of the North Kimberley.[22]

The surveys of the Prince Regent River Reserve and Drysdale River National Park in 1974–75 demonstrated that the monsoon vine thickets are not an isolated phenomenon of Mitchell Plateau. Aerial inspection from flights to and from the Plateau showed that monsoon vine thicket patches are common throughout the north-west coastal region, and subsequent visits by sea to this area confirmed their

FIGURE 23

Karl Pirkopf (field assistant), Kevin Kenneally (botanist) and Roger Hnatiuk (ecologist) at the Amax campsite, Mitchell Plateau in June 1976.

FIGURE 24

Kevin Kenneally and Daphne Edinger collecting rainforest plants at Sale River in June 1986. *Image: Kevin Coate*

occurrence at the Walcott Inlet, Sale and Glenelg Rivers, Camden Sound, and St George Basin on the Prince Regent River.

The known range of the monsoon vine thickets was extended to the south-west Kimberley, an area of substantially lower rainfall, when a biological survey of the Dampier Peninsula documented their occurrence behind coastal dunes, as foreshadowed by Brockman in 1880. The thickets occur in frequent patches, more commonly at the north end of the Peninsula, on the lee side of the coastal dune system. This habitat was not known to be occupied by monsoon vine thickets in Western Australia, though it is a common element in the Northern Territory and Queensland.[23]

Between 1987 and 1989, the first broad-scale quantitative ecological survey of the Kimberley rainforests was undertaken.[24] Data were collected on compositional patterns of biota, disturbance and physical characteristics of the environment at 95 rainforest patches. The results of this survey provided information on soil and landform, invertebrates and vertebrate fauna, floristics, vegetation structure and spatial distribution of rainforest patches.

In 1998, INPEX Corporation was awarded a petroleum exploration permit in the northern Browse Basin, about 200 km north-west of the Kimberley coast. The company carried out biodiversity inventories and habitat surveys of the basin and selected islands.[25] Environmental studies on the Maret Islands confirmed the presence of rainforest (coastal vine thicket) patches on the slopes around the edge of the laterite plateau. They also established that the extent of contiguous rainforest (vine thicket) on South Maret Island made it one of the larger tracts of intact vine thickets in the Kimberley. Preliminary observations of the vine thickets on sandstone islands in the region indicated that they contained species assemblages and structures that differed from those of the vine thickets on lateritic islands. The studies in the Browse Basin were completed in 2008, but were no longer required when INPEX chose Darwin as the preferred site for its gas plant.

Island populations play an important role in maintaining biodiversity over geological time. As plant populations on islands in the Kimberley have been isolated from the mainland for up to thousands of years, they could represent a unique reservoir of useful genetic information.[26,27] Between 2007 and 2010, the Western Australian Department of Environment and Conservation carried out a systematic survey of plants and animals on 24 selected inshore islands off the northern Kimberley coast. These surveys showed that the largest and wettest islands supported well-developed rainforest patches.[28]

Karl Lagoon, upper Calder River on the Munja Track.

CHAPTER 5
Climate and weather in the Kimberley

The climate of the Kimberley is dominated by the Australian summer monsoon, which forms part of a wider climate system that includes the Indo-Pacific Warm Pool. This pool of warm ocean water is a global heat source that has a major role in driving planetary-scale circulation.[1,2] The tropical monsoon climate is characterised by two distinct seasons, a 'wet' and a 'dry', with several transition periods within each season (Fig. 25). Precipitation in the Kimberley is driven by the summer monsoon and tropical cyclones, both of which typically occur during the austral summer between November and March and are associated with the seasonal southward migration of the Inter-Tropical Convergence Zone.[3] Generally, the wet season lasts from November to March and the dry season from April to October. The dry season is virtually rainless, with easterly winds and cooler nights. The wet season, however, can bring torrential

FIGURE 25

Cumulonimbus storm clouds building up over the Amax Bauxite Corporation campsite at Mitchell Plateau at the onset of the wet season in 1970. *Image: Joe Smith*

rain and widespread flooding associated with cyclones and tropical thunderstorms (Fig. 26). The average annual rainfall for the region is 950 mm, with Mitchell Plateau receiving an annual average of 1500 mm. Temperatures are high throughout the year, with annual averages between 25°C and 35°C.[4]

Monsoon rainforest in the Kimberley is present only in small scattered patches within 150 km of the north-west coast, extending from Broome in the south, where the minimum annual rainfall is 600 mm, to Cape Londonderry in the north, where the maximum is 1500 mm. Outlying peninsulas and islands are relatively dry, with the highest rainfall occurring on elevated ground inland of the main coastline.

The patches of monsoon vine thickets on the Dampier Peninsula, in contrast to other Kimberley and northern Australian rainforests, are at the southern limit of their range, occur in the swales of coastal dunes, and are dependent on groundwater from the Broome aquifer.[5] Monsoon forests are not known to occur where the total annual rainfall is below 600 mm. On average, Broome receives 540 mm, Beagle Bay 725 mm and Cape Leveque 718 mm for these locations, trending northward on the Dampier Peninsula. The survival of these coastal patches may be aided by fog, which provides supplementary water (Fig. 27). On average, Broome (and the Dampier Peninsula) has 22 foggy days each year from June to September – the dry season.[6] Fog can persist all day. In July 1986, for example, I observed fog at Middle Lagoon, 25 km north-west of Beagle Bay, that was so dense, visibility was reduced and the vegetation was dripping wet. The fog extended 19 km offshore to the Lacepede Islands. The same has been observed at Mitchell Plateau in June (Fig. 28). Fog is an important source of water for plants and animals in arid country, which has been documented for the Namib Desert in southern Africa.[7,8]

Over the past 10,000 to 20,000 years, the climate of the Kimberley changed, often abruptly, in ways that would have required versatile responses from Aboriginal people.[9] Researchers have also discovered evidence of a severe drought in the region between 1000 and 2000 years ago, which would have had a profound impact on its inhabitants.

Other evidence suggests abrupt interruptions to the northern Kimberley monsoon.[9] These interruptions were driven primarily by the collapse of sea ice into the Atlantic Ocean, which in turn affected weather in East Asia, itself linked to the life-giving rains. This research also supports the notion that people likely made early summers hotter and drier by burning the landscape at the end of winter to facilitate hunting and to promote the survival of specific plants and animals.

FIGURE 26
(LEFT)

Mitchell River and Mitchell Falls after Tropical Cyclone Bruno in January 1982.

FIGURE 27
(BELOW LEFT)

Coastal fog, with its saturated moist air, extending from Middle Lagoon Resort on the west coast of the Dampier Peninsula to the Lacepede Islands, 19 km offshore, in June 1987.

FIGURE 28
(BELOW)

Fog at the Amax campsite at Mitchell Plateau in June 1982.

Two primary components of tropical precipitation – monsoons (tropical lows) and tropical cyclones – can provide large volumes of rainfall in short periods of time, which leads to flooding. Because both systems respond to changes in atmospheric and sea surface conditions, it is imperative that we understand their sensitivities to climate change.[10]

There has been little effort to monitor the impact of climate change on either the Kimberley or its unique monsoon rainforests and surrounding vegetation. Climate-change scenarios suggest that a number of threats are imminent, including higher temperatures, more frequent fires and extreme events, and the loss of endemic plant and animal species.[11] An increase in the number of extreme heat events in the Kimberley will have severe effects on the wellbeing of people in the region, particularly Indigenous communities. As well as affecting key industries, including tourism and agriculture, changed temperature and rainfall regimes will damage social and natural ecosystems. The number of days over 40°C in Derby is expected to increase dramatically in the coming decades, according to the eight climate models used by CSIRO and the Bureau of Meteorology. Under a business-as-usual scenario for greenhouse gas emissions, CSIRO projects that Derby could experience as many as 41 days over 40°C per year by 2030, and up to 168 days per year by 2090. The average from 1981 to 2010 was only 14 days per year.[11]

FIGURE 29   Wanjina cloud over Port George IV, directly east of Augustus Island. Each wet season, the air becomes thick with electricity, banks of cumulonimbus clouds form, and everywhere the power of the Wanjina can be seen, felt and heard. *Image: Tim Willing*

FIGURE 30

Sandstone gallery showing a painted Wanjina, the main creation spirit associated with rain and seasonal regeneration. The body of a Wanjina is often shown covered with dots, which represent rainfall. *Image: Wunambal Gaambera Aboriginal Corporation*

INDIGENOUS KNOWLEDGE OF SEASONS

Aboriginal concepts of seasonality are more complex than that of a wet-dry season pattern. The Bardi and Yawuru peoples recognise six seasons, distinguished mainly by wind, rainfall direction and intensity, ripening of fruits, and the appearance or disappearance and 'fatness' of fish and animals.[12] The Wunambal Gaambera people of the North Kimberley recognise the onset and duration of four major seasons: *Wunju* (wet season), *Bandemanya* (early dry season), *Yurrma* (cold dry season) and *Yuwala* (build-up to wet season).[13]

In the North Kimberley, rainfall and seasonal regeneration were often directly attributed to the sacred Wanjina (also spelled Wandjina and Wondjina) (Fig. 29).

Images of Wanjinas, which are found in rock shelters and sandstone galleries in the Kimberley, show a shape-changing, anthropomorphic being in a full-frontal pose, with prominent eyes but no mouth (Fig. 30). They are often depicted with elaborate headdresses that may indicate different types of storms. The body of a Wanjina is often shown covered with dots that represent rainfall.[14] These images were retouched to ensure the coming of the monsoon and its regeneration of all life.[15]

Porosus Creek and Naturalists Island, Prince Frederick Harbour. *Image: Graeme Snow/Shutterstock*

CHAPTER 6
Monsoon rainforest environments

The landscape in the north-west Kimberley is underlain mostly by sandstone but also partly by basalt of Precambrian age. The deeply weathered sandstone country is generally very rugged, consisting of a high degree of stacks of bare rock and extensive pavements often covered in hummock grasslands (*Triodia* species). Near the coast, the country is more rugged and barren, but further inland, it is more even, undulating and soil covered. Sandstone country is typically dissected by narrow gorges formed along joints in the rock, thus creating a rectilinear drainage pattern (Fig. 31).

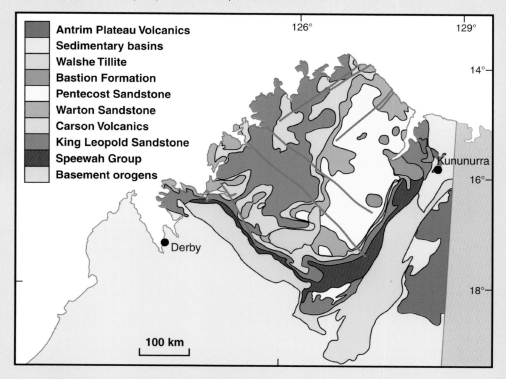

Geology of the Kimberley. Reproduced with permission from Mike Donaldson.[1]   FIGURE 31

Antrim Plateau Volcanics
Sedimentary basins
Walshe Tillite
Bastion Formation
Pentecost Sandstone
Warton Sandstone
Carson Volcanics
King Leopold Sandstone
Speewah Group
Basement orogens

126°   129°

14°

Kununurra

16°

Derby

18°

100 km

FIGURE 32

A SIMPLIFIED GUIDE TO THE TROPICAL MONSOON RAINFORESTS AND VINE THICKETS OF KIMBERLEY, WA

COASTAL SAND DUNES

MARET ISLANDS

DAMPIER PENINSULA

DAMPIER PENINSULA

DAMPIER PENINSULA

DAMPIER PENINSULA

SAND DUNE FRINGING ARAFURA SEA

HILLSIDES AND SCREE SLOPES

NATURALISTS ISLAND

CAPE BOUGAINVILLE

MITCHELL PLATEAU

MITCHELL PLATEAU

MITCHELL PLATEAU

PORT WARRENDER

SWAMPS AND RIVERS

WALCOTT INLET

WALCOTT INLET

MITCHELL PLATEAU

MITCHELL PLATEAU

ANJO PENINSULA

SPECIES POOR GORGES GULLIES AND SLOPES

LIMESTONE RANGES, NORTH KUNUNURRA

MANGROVES

NORTH KIMBERLEY

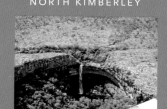

NINGBING RANGE

RHIZOPHORA STYLOSA

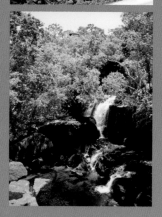

SONNERATIA ALBA

EAST KIMBERLEY

SONNERATIA ALBA

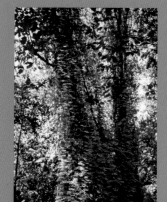

EL QUESTRO

POROSUS CREEK

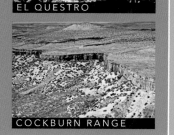

COCKBURN RANGE

WALSH POINT

FIGURE 33 Patch of monsoon rainforest known as Crusher Thicket on Mitchell Plateau.

Basalt country, by contrast, typically consists of rounded hills of moderate to low relief, and richer, deeper soils. Outcrops of boulders and sheet rock are common. Near the coast, relief is greater, and dissection of individual lava flows from the basaltic ground mass gives rise to a topography of pronounced structural benches and mesas, as well as a dendritic drainage pattern. In the North Kimberley, the Carson Volcanics overlie King Leopold Sandstone at thicknesses that generally range between 400 and 700 metres but can reach 1100 metres.[2]

There is very little soil cover in the Kimberley, except in some low-lying areas, with vegetation growing in crevices and cracks in the sandstone. Rainforest patches occur on a wide variety of soils, from siliceous sands to red clays, although within the patches soils are generally uniform (Fig. 32). A study of soils of adjacent savanna woodlands found they were similar to those within the rainforest patches.[3] Soil chemical analysis found generally low levels of nutrients at the surface, with nutrient concentrations increasing with soil depth. Rainforest soils tended to have higher nutrient levels than soils of the adjacent woodlands. No clear explanation for the presence of rainforest patches has been ascertained from soil and landform data.[3]

FIGURE 35 (ABOVE)

Monsoon rainforest (*Wulo*) on a scree slope on Naturalists Island at the mouth of the Hunter River, Kimberley.

FIGURE 36 (LEFT)

Monsoon rainforest fringing the dissected laterite plateau at Cape Bougainville.

FIGURE 37 (BELOW)

Riparian forest dominated by paperbarks (*Melaleuca leucadendra*) and Screwpines (*Pandanus aquaticus* and *P. spiralis*).

FIGURE 34 (TOP)

Isolated patch of monsoon rainforest on Mitchell Plateau surrounded by savanna woodland.

FIGURE 38
(TOP LEFT)

Monsoon rainforest on Carson Volcanic benches at Port Warrender, Mitchell Plateau. *Image: Tim Willing*

FIGURE 39
(TOP RIGHT)

Tony Raudino, a member of the 1993 Landscope Expedition to Mitchell Plateau, standing on a fallen tree festooned with vines at the Lone Dingo rainforest patch.

FIGURE 40
(CENTRE LEFT)

Joe Smith, the manager of the Amax mining campsite, and Karl Pirkopf, a field assistant to Kevin Kenneally, exploring a rainforest patch in a gorge on the edge of Mitchell Plateau in 1978.

FIGURE 41
(ABOVE RIGHT)

Dr Bernie Hyland and Kevin Kenneally collecting plant specimens in a rainforest patch on South West Osborne Island during a Kimberley rainforest survey in 1987.

FIGURE 42
(ABOVE)

Rainforest on a levee bank on the King Edward River, south of Kalumburu. *Image: Norm McKenzie*

Tall closed-canopy swamp rainforest on the edge of a tidal mudflat at Walcott Inlet. *Image: Norm McKenzie*

FIGURE 43
(TOP LEFT)

Tidal creek at the north end of Porosus Creek, a tributary of the Hunter River, showing the close proximity of monsoon rainforest (fringing the base of the sandstone cliffs) to mangroves.

FIGURE 44
(ABOVE RIGHT)

Patchy monsoon rainforest on exposed scree slopes in the Cockburn Range, East Kimberley.

FIGURE 45
(ABOVE LEFT)

Basalt country features remnants of a duricrusted plateau surface, the largest of which are Mitchell Plateau (with a surface area of 220 square kilometres) and the Bougainville Peninsula (85 square kilometres).[4] The duricrust is bauxite, some 2 metres thick, and underlain by a kaolinised weathered zone averaging 17 metres in thickness. These remnants take the form of conspicuous mesas bordered by deeply indented scarps.

On the Bougainville Peninsula, monsoon rainforest is widespread on the undulating basalt country and may form semi-deciduous monsoon rainforest on scree slopes of the plateau. In both cases, *Eucalyptus* and *Corymbia* savanna woodland occurs either as pure patches or mixed with monsoon vine thickets (Figs 33–45).

FIGURE 46   The southern-most patch of monsoon rainforest (vine thicket)
             behind Holocene coastal dunes at Coulomb Point, north
             of Broome.

Botanical exploration of the north-west coastal country between King
Sound and the Admiralty Gulf has found that the vegetation typical
of the Mitchell Plateau area is repeated, with patches of monsoon vine
thicket and forest on benched basalt topography and on sandstone
scree slopes overlying basalt at the foot of sandstone scarps. This latter
habitat recalls the breakaway slope habitat of Mitchell Plateau and is
developed where overlying Warton and King Leopold Sandstone has
exposed underlying basalt, a common situation between Doubtful
Bay and Prince Frederick Harbour. The occurrence of monsoon vine
thicket is by no means confined to the basalt country, but occurs less
frequently on the sandstone.[5]

Monsoon rainforest on coastal dunes near Broome is dominated in the wet season by Crab's Eye Bean (*Abrus precatorius*), which bears distinctive, shiny, scarlet-red seeds that are highly toxic to humans.

FIGURE 47

Further south, an entirely different habitat is found on the Dampier Peninsula, where the average annual rainfall ranges from 600 to 800 mm. The substrate is deep Holocene dune sands, white except for a superficial dark grey organic layer covered by up to 6 cm of leaf litter.[6] Vegetation includes vine thickets and associated plant species from coastal sand dunes (Figs 46, 47). This habitat has not been encountered elsewhere in the north-west Kimberley, except behind the beaches on some islands (for example, the Maret Islands).

A buttressed rainforest tree on the island of Wetar (part of Wallacea), Indonesia. *Image: Ric How*

CHAPTER 7
The biocomplexity of monsoon rainforests

Four major categories of monsoon rainforest have been recognised in the Kimberley: those on hillsides and scree slopes; along swamps and rivers; behind Holocene coastal sand dunes; and in gorges and gullies.[1,2] In addition, there are patches containing a limited number or range of rainforest species on the Devonian limestones of the Ningbing Ranges north of Kununurra.[i]

Mangrove communities of South-East Asia have been treated as a special sort of rainforest, because they include not only true mangrove species but also plant species normally found in other habitats.[3,4] Kimberley mangrove communities are completely distinct floristically from rainforest, but because they are often contiguous or in close proximity to rainforest patches and have served as refugia during sea-level changes, they are considered as an extension of the closed-canopy community. Many of the bird and mammal species of the Kimberley rainforest also depend on the vast stands of mangroves that fringe the tidal creeks and coastline (Fig. 48).

South-East Asia, Australia and the landmasses between them are home to some of the most unique, diverse and threatened flora in the world.[5] This region, a hotspot for biogeographical research, is characterised by the emergence of the Sunda and Sahul continental shelves. The fact that plant lineages were exchanged between Australia and South-East Asia is well established; however, how this floristic exchange has influenced plant geography across the region is not yet well understood.[6]

---

i. This information was provided by Norm McKenzie in a personal communication.

FIGURE 48   Diagram of the structure of Kimberley monsoon rainforests.
             Modified with permission from the Wet Tropics Management Authority.

FIGURE 49   A geological time scale of Earth based on stratigraphy,
             which is a correlation and classification of rock structure.

EON	ERA	PERIOD		EPOCH	
Phanerozoic	Cenozoic	Quaternary		Holocene	← TODAY
				Pleistocene	← 11.8 Ka
		Neogene		Pliocene	
				Miocene	
		Paleogene		Oligocene	
				Eocene	
				Paleocene	← 66 Ma
	Mesozoic	Cretaceous		-	
		Jurassic		-	
		Triassic		-	← 252 Ma
	Paleozoic	Permian		-	
		Carboniferous	Pennsylvanian	-	
			Mississippian	-	
		Devonian		-	
		Silurian		-	
		Ordovician		-	
		Cambrian		-	← 541 Ma
Proterozoic	-	-		-	← 2.5 Ga
Archean	-	-		-	← 4.0 Ga
Hadean	-	-		-	← 4.54 Ga

Younger ↑ Older

Australian rainforest communities are believed to contain plant and animal species with origins in Gondwana and Laurasia - the two supercontinents that formed Pangaea.[1] They should be seen as being composed of indigenous albeit ancestral species, in which case rainforests can be seen as relictual pockets of ancestral vegetation.[7]

About 49 million years ago, the Australian Plate separated from the Antarctic continent and began drifting northward,[8,9] carrying with it species of rainforest vegetation that were indigenous to Gondwana. On its journey north, the Australian continent escaped the ravages of Antarctic freezing, only to heat up and dry out as it approached the tropics. As the climate changed, the rainforests contracted, eventually covering only 1% of the landmass by the time of European settlement (Fig. 49).

Following the collision of the Australian Plate with Asia during the Miocene (the period covering ~23–5 million years ago), numerous rainforest plant species migrated into northern Australia.[5,6] Fossil

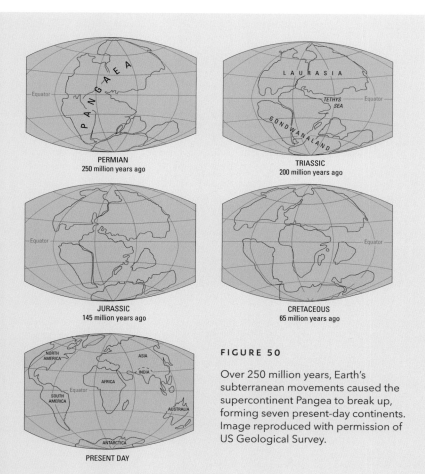

PERMIAN
250 million years ago

TRIASSIC
200 million years ago

JURASSIC
145 million years ago

CRETACEOUS
65 million years ago

PRESENT DAY

FIGURE 50

Over 250 million years, Earth's subterranean movements caused the supercontinent Pangea to break up, forming seven present-day continents. Image reproduced with permission of US Geological Survey.

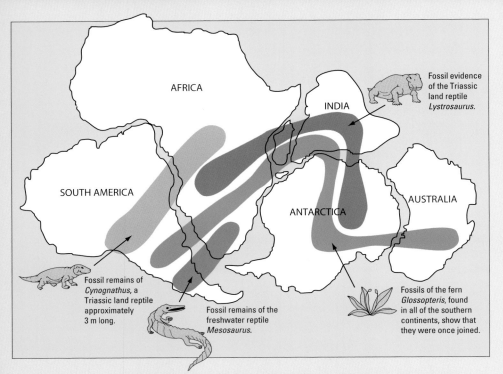

Fossil evidence of the Triassic land reptile *Lystrosaurus*.

AFRICA

INDIA

SOUTH AMERICA

AUSTRALIA

ANTARCTICA

Fossil remains of *Cynognathus*, a Triassic land reptile approximately 3 m long.

Fossil remains of the freshwater reptile *Mesosaurus*.

Fossils of the fern *Glossopteris*, found in all of the southern continents, show that they were once joined.

FIGURE 51   Continental drift explains why look-alike animal and plant fossils, and similar rock formations, are found on different continents. Image reproduced with permission of US Geological Survey.

evidence suggests that most of the Australian continent in the Tertiary (65–1.8 million years ago) was covered by rainforest. Rainforests were widespread in Australia in the late Eocene (35 million years ago), including in inland areas that are now arid.[10] Palaeobotanical and phylogenetic analyses have shown that floristic exchange between Australia and South-East Asia began around 30 million years ago.[11] This coincided with the increasing connectivity of Australia and South-East Asia as the continental (Sahul) shelf on which Australia is situated began to collide with the Sunda Shelf (Figs 50, 51).[12]

Increasing aridity also led to the loss of rainforests in central Australia around 10 million years ago.[13] Indomalayan rainforest species (the more intrusive species) began colonising the northern part of the Australian Plate from the early Tertiary,[14] and "a strong contact between the floras of the Australian and Indomalayan regions... established from Miocene times".[15] The convergence of the Sunda Shelf and Sahul Shelf in the Miocene facilitated the exchange of previously isolated floras between these two regions.[16] Over the past 20 million years (especially during the Miocene and the Pliocene, 5.3–2.6 million years ago), as climates became more variable and perhaps seasonally arid,[17] these rainforests were gradually replaced by more open scleromorphic and xeromorphic vegetation.[18] Rainforests similar to those that once covered the ancient supercontinent of Gondwana are now restricted to the wet tropics of north-east Queensland, especially Daintree National Park.[19,20]

During the Pleistocene Epoch (2.6 million to 11,700 years ago), Kimberley rainforest species may have survived fluctuations in mangrove or littoral vegetation near the edge of the exposed continental shelf, in riverine vegetation, or in refuges associated with the Kimberley Plateau.[21] Studies of pollen, mangrove clays, woods and organic muds from the North Kimberley suggest that extensive mangrove forests were established at least 9000 years ago and were sustained until at least 6000 years ago, facilitated by rising sea level, increasing postglacial temperatures and greater amounts of fresh water from a more active monsoon.[22-24] The study of pollen from the region's north also revealed woodland expansion at this time; the presence of species that depend on moist soils suggests that moisture was increasing.[24]

At Black Springs, an organic mound spring in the north-west Kimberley, the Pleistocene-Holocene Transition (which ended about 11,000 years ago) was a period of significant climatic and environmental change.[25] Records indicate that since this transition, the Australian summer monsoon has varied greatly in intensity, with an increase in monsoonal precipitation from around 14,000 years ago,[26] followed by pronounced drying late in the Holocene (a period covering the last few thousand years). In general, conditions in tropical Australia were wetter and warmer in the early Holocene than previously. Speleothem records from the south-west and eastern Kimberley also indicate that monsoon conditions in the early Holocene were wetter than previously experienced.[27,28] When local conditions became more mesic – featuring moderate supplies of moisture – during the late Pleistocene and early Holocene, rainforest species may have re-colonised the region from the Northern Territory, Queensland or Indonesian islands to the north via dispersal vectors such as frugivorous (fruit-eating) birds and bats. During the late Holocene, a less active monsoon may have caused rainforests to contract.

RAINFOREST PLANTS

Trees and shrubs

Because their seeds are transportable over vast distances, rainforest plants can readily invade areas of suitable habitat. By maintaining their population in isolated patches and even colonising new patches, these plants provide additional habitats for animals.

FIGURE 52 (TOP LEFT)

The Droopy Leaf (*Aglaia elaeagnoidea*) occurs in the Kimberley, Cape York Peninsula and elsewhere in Queensland, as well as in India, Asia, Malesia, Taiwan and New Caledonia.

FIGURE 53 (TOP RIGHT)

The fruits of the Droopy Leaf.

FIGURE 54 (ABOVE LEFT)

Mamajen or *Walara* (*Mimusops elengi*) is widespread in monsoon rainforest across northern Australia.

FIGURE 55 (ABOVE)

The Canary Beech (*Monoon australe*) is endemic to Australia and widespread across the tropical north.

FIGURE 56 (BELOW)

The Banyan Fig or *Yawurru* (*Ficus virens*) has aerial and strangling roots. *Image: Tim Willing*

Monsoon rainforest patches in the North Kimberley are typified by tall emergent crowns of tree species that reach more than 15 metres in height. These species include the Droopy Leaf (*Aglaia elaeagnoidea*), Alligator Bark (*Calophyllum sil*), Canarium (*Canarium australianum*), Brittlewood (*Claoxylon hillii*), Cunningham's Laurel (*Cryptocarya cunninghamii*), *Dysoxylum latifolium*, False Olive (*Elaeodendron melanocarpum*), Scaly Ash (*Ganophyllum falcatum*), Mamajen (*Mimusops elengi*), Canary Beech (*Monoon australe*) and Australian Nutmeg (*Myristica insipida*) (Figs 52–55). They also include species that are deciduous or semi-deciduous during the dry season: the Siris Tree (*Albizia lebbeck*), Kurrajongs (*Brachychiton diversifolius, B. xanthophyllus*), Kapok Tree (*Bombax ceiba*), Ficus species (*F. congesta, F. geniculata* var. *insignis, F. hispida* var. *hispida, F. racemosa* var. *racemosa* and *F. virens* var. *virens*), Garuga (*Garuga floribunda*), Goolmi or Coffee Bush (*Grewia breviflora*), Brown Bollywood (*Litsea glutinosa*), *Miliusa brahei*, Peanut Tree (*Sterculia quadrifida*), *Terminalia* species (*T. ferdinandiana, T. microcarpa, T. petiolaris* and *T. sericocarpa*), *Vavaea amicorum, Wrightia pubescens* and Ziziphus (*Ziziphus quadrilocularis*) (Figs 56–59).

FIGURE 57 (ABOVE)

Loofah Vine (*Luffa saccata*) growing on *Marool* or *Larriya* (*Terminalia petiolaris*). *Image: Tim Willing*

FIGURE 58 (ABOVE RIGHT)

The Banyan Fig is widespread in monsoon rainforests across northern Australia, Asia and Malesia.

FIGURE 59 (RIGHT)

The Loofah Vine occurs in Western Australia and the Northern Territory. *Image: Tim Willing*

The profile of the canopy is characteristically irregular in height and appears to consist of vine foliage as much as tree foliage. The understorey is dominated by small trees and shrubs, such as the Wild Randa (*Aidia racemosa*), *Aglaia brownii*, *Alectryon* species (*A. connatus* and *A. kimberleyanus*), Great Woolly Malayan Lilac (*Callicarpa candicans*), Goonj (*Celtis strychnoides*), Lolly Bush (*Clerodendrum floribundum*), *Denhamia obscura*, Broad Leaved Ebony (*Diospyros maritima*), Yellow Tulipwood (*Drypetes deplanchei*), *Gooralgarr* or Snowball Bush (*Flueggea virosa* subsp. *melanthesoides*), *Glycosmis* species (*G. macrophylla* and *G. trifoliata*), *Grewia glabra*, *Mallotus dispersus*, Fingersop (*Meiogyne cylindrocarpa*), *Memecylon pauciflorum*, Lime Berry (*Micromelum minutum*), Orange Jasmine (*Murraya paniculata*), Shiny-leaved Canthium (*Psydrax odorata* subsp. *arnhemica*), Strychnine Bush (*Strychnos lucida*), Banana Bush (*Tabernaemontana orientalis*), Tree Ixora (*Tarenna dallachiana*), and Peach-leaved Poison Bush (*Trema tomentosa*) (Figs 60-62).

On the rainforest floor, there is virtually no flora apart from seedlings. Two unusual plants recorded from the rainforest floor are the Coastal Rein Orchid (*Habenaria hymenophylla*) and the pan-tropical Arrowroot (*Tacca leontopetaloides*). The Arrowroot grows to one metre and produces deeply divided leaves and flowers on separate stalks; the flowers open at night. Aboriginal people eat the tuber of this plant after extensive preparation. Other species that have been recorded from the rainforest floor include the pan-tropical Chaff Flower (*Achyranthes aspera*), Water Nutgrass (*Cyperus aquatilis*), Musk-scented Plant (*Hypoestes floribunda* var. *suaveolens*) and Forest Burr (*Pupalia lappacea*).

Species commonly found along the edges of monsoon rainforest patches include the wattles (*Acacia aulacocarpa*, *A. holosericea* and *A. pellita*), *Brachychiton* species (*B. tridentatus* and *B. viridiflorus*), the Broad-leaved Bloodwood (*Corymbia foelscheana*), Boab (*Adansonia gregorii*), *Bilanggamar* or Helicopter Tree (*Gyrocarpus americanus*), White Cedar (*Melia azedarach*) and Black Currant Tree (*Antidesma ghaesembilla*).

Lianas and vines

Creepers, vines and lianas (woody vines) are abundant in the canopy and make up a significant proportion of rainforest vegetation. Despite their importance to the rainforest flora, the ecology of vines and lianas in Kimberley monsoon rainforests is poorly understood.

Vines form impenetrable tangles, and many plants have adapted to climb into the canopy. Modifications include prickles, spines or hooks, such as those found on *Asparagus racemosus*, Yellow Nicker

The *Goonj* or Nettle Tree (*Celtis strychnoides*) is widespread in northern Australia and also found in the Americas, Madagascar, Malesia and Pacific Islands.

The Broad Leaved Ebony (*Diospyros maritima*) is widespread across northern Australia, Asia and New Guinea. *Image: Tim Willing*

Moolinyj (*Glycosmis trifoliata*) is widespread across northern Australia, Asia and Malesia.

(*Caesalpinia major*), bush capers (*Capparis lasiantha* and *C. quiniflora*), *Luvunga monophylla*, Spitting Cucumber (*Muellerargia timorensis*), Thorny Pisonia (*Pisonia aculeata*), Climbing Wattle (*Senegalia albizioides*) and Lawyer Vine (*Smilax australis*) (Figs 63–65). Some species use tendrils, such as Lacewing Vine (*Adenia heterophylla*), Native Grape (*Ampelocissus acetosa*), Coastal Water Vine (*Causonis maritima*) and Three-leaf Cayratia (*Causonis trifolia*), the native grapes *Cissus adnate* and Slender Water Vine (*Cissus reniformis*), Striped Cucumber (*Diplocyclos palmatus*) and the gourds *Trichosanthes cucumerina* and *T. pilosa*).

FIGURE 63 (ABOVE)

In Australia, *Goolyi* or Yellow Nicker (*Caesalpinia major*) is found only in the Kimberley. It has also been recorded in the Americas, Madagascar, Asia, Malesia and the Pacific Islands.

FIGURE 64 (LEFT)

The seed pods of *Goolyi*.

FIGURE 65 (BELOW LEFT)

The Climbing Wattle (*Senegalia albizioides*) has been recorded in Western Australia and in Cape York Peninsula, Queensland. *Image: Tim Willing*

FIGURE 66 (LEFT, FROM TOP TO BOTTOM)

Gaaji, *Balbal*, Supplejack or Bush Cane (*Flagellaria indica*) is a vigorous cane-like creeper with fragrant cream flowers. It is widespread across northern Australia, Asia, Malesia and the Pacific Islands.

FIGURE 67

Bush Cane fruits are red, globular and inedible.

FIGURE 68

Goolmi, *Yan gai*, Currant or Coffee Bush (*Grewia breviflora*) is widespread in the Kimberley and also found in the Northern Territory, Cape York Peninsula, Timor and New Guinea.

FIGURE 69

Fruits of the *Goolmi* can be eaten raw when ripe.

FIGURE 70 (BELOW)

The Black Bean (*Mucuna gigantea*) is found across northern Australia and New South Wales.

Other species have modified tendrils, including Tropical Clematis (*Clematis pickeringii*) and Bush Cane (*Flagellaria indica*) (Figs 66, 67). Modifications such as stem twiners or scramblers are found on Crab's Eye Bean (*Abrus precatorius*), Dutchman's Pipe (*Aristolochia acuminata*), Wild Jack Bean (*Canavalia papuana*), *Yugulu* or Love Vine (*Cassytha filiformis*), Milk Pods (*Cynanchum carnosum* and *C. pedunculatum*), *Goolmi* or Coffee Bush (*Grewia breviflora*), *Grewia guazumifolia*, Blush Plum (*Grewia oxyphylla*) and Emu-Berry (*Grewia retusifolia*), Sweet Morinda (*Gynochthodes jasminoides*), Mauve Clustervine (*Jacquemontia paniculata*), Native Jasmine (*Jasminum didymium*), Fingersop (*Meiogyne cylindrocarpa*), Raspberry Jelly Tree (*Miliusa brahei*), Itchy Bean (*Mucuna diabolica* subsp. *kenneallyi*), Black Bean (*Mucuna gigantea*) and Creeping Bean (*Mucuna reptans*), *Operculina codonantha*, *Pachygone ovata*, *Thunbergia arnhemica*, *Waramburr* or Snake Vine (*Tinospora smilacina*) and Slender Cucumber (*Zehneria mucronata*) (Figs 68–70). Other adaptations include pods with irritant hairs (such as those of the Itchy Bean) and stems that emit a milky exudate, such as that of the Fire Vine (*Trophis scandens*), which can cause painful 'burns' or dermatitis on contact with skin.

During the wet season, the abundant vine foliage envelops the tree canopy, reducing the amount of sunlight that can penetrate and increasing the humidity below. Vines and lianas often begin life on the ground as small self-supporting shrubs and rely on other plants to reach the light-rich environment of the upper canopy. Upon reaching the canopy, vines and lianas spread from tree to tree. This creates important pathways for canopy-dwelling animals, which feed on the abundant leaves, flowers and bright, fleshy fruits - a distinctive feature of rainforests in the wet tropics. The flowers are also important for pollinators (Figs 71–98).

FIGURE 71 (OPPOSITE)

The hemiparasitic *Jarnba* or Mistletoe Tree (*Exocarpos latifolius*) is widespread across northern Australia and Malesia.

FIGURE 72 (TOP)

The fruits of *Jarnba* have a swollen stalk resembling a cultivated cashew. When ripe, the stalk is edible but relatively tasteless.

FIGURE 73 (ABOVE)

Euphorbia plumerioides occurs in Western Australia, Cape York Peninsula and North-East Queensland, as well as in Asia and Malesia.

FIGURE 74 (MIDDLE ABOVE)

Marool or Blackberry Tree (*Terminalia petiolaris*) is endemic to coastal areas of the Kimberley.

FIGURE 75 (ABOVE)

The Drypetes or Yellow Tulipwood (*Drypetes deplanchei*) is a small to medium-size tree that grows up to 25 metres in height. It is widespread across northern Australia, with a range that extends into New Guinea and New Caledonia.

FIGURE 76 (TOP LEFT)

Arndany or Asparagus Fern (*Asparagus racemosus*) is widespread across northern Australia, Asia and Malesia.

FIGURE 77 (LEFT)

Wunbula, *Wangula* or Lacewing Vine (*Adenia heterophylla*) is widespread across northern Australia and Malesia.

FIGURE 78 (TOP)

Waramburr or Snake Vine (*Tinospora smilacina*) is endemic to Australia and widespread across northern Australia.

FIGURE 79 (ABOVE)

The Indian Prickly Ash (*Zanthoxylum rhetsa*) occurs in Western Australia, Cape York Peninsula and North-East Queensland, as well as in Asia and Malesia.

FIGURE 80 (TOP)

Flowers and fruits of the *Ngaming* or Crab's Eye Bean (*Abrus precatorius*), which is widespread across northern Australia, Africa, Asia, Malesia and Pacific Islands. *Image: Brian Carter*

FIGURE 81 (ABOVE)

The *Gubinge, Arungul* or *Madoor* (*Terminalia ferdinandiana*) is a deciduous tree with white flowers that have a sweet nectar smell. It is widespread across northern Australia.

FIGURE 82 (ABOVE)

The large fleshy fruits of *Gubinge*, which are whitish green and sometimes tinged with pink, are eaten raw when ripe. The main fruiting season is January to February, depending on rainfall.

FIGURE 83 (TOP LEFT)

The Cluster Fig, River Fig, *Jiliwa* or *Ganjirr* (*Ficus racemosa*) is widespread across northern Australia, Asia and Malesia.

FIGURE 84 (BOTTOM LEFT)

The Lolly Bush (*Clerodendrum floribundum*) is widespread across northern Australia, Timor and New Guinea.

FIGURE 85 (TOP RIGHT)

The Itchy Bean (*Mucuna diabolica* subsp. *kenneallyi*) is endemic to Kimberley monsoon rainforests.

FIGURE 86 (BOTTOM RIGHT)

The seed of Australian Nutmeg (*Myristica insipida*) is covered by a reddish lace-like appendage, or aril. The fruit is a favoured food of the Pied Imperial Pigeon. The species is widespread across northern Australia, Malesia and the Solomon Islands.

FIGURE 87 (OPPOSITE)

Gooralgarr, *Arnbarrma* or Snowball Bush (*Flueggea virosa* subsp. *melanthesoides*) is widespread across northern Australia and New Guinea.

FIGURE 88 (LEFT)

The tree *Melochia umbellata* is restricted in Australia to Kimberley monsoon rainforests, but is also found in South-East Asia.

FIGURE 89 (MIDDLE LEFT)

Alectryon kimberleyanus is a small tree restricted to Kimberley monsoon rainforests.

FIGURE 90 (BELOW LEFT)

Trichosanthes cucumerina var. *cucumerina*, a vine belonging to the cucumber family, is widely distributed in India, Southern China, South-East Asia, Malesia and northern Australia.

FIGURE 91 (BELOW)

The papery wings of the Stinkwood cause the fruit to spin, allowing it to fly considerable distances.

FIGURE 92

The *Bilanggamar*, *Minda*, Helicopter Tree or Stinkwood (*Gyrocarpus americanus*) is widespread within and on the edges of monsoon rainforest across northern Australia. It also occurs in East Africa, India, Malesia, Melanesia, Polynesia and Central and South America.

FIGURE 93   *Kulalart* or Creeping Bean (*Mucuna reptans*), a twining or climbing sea-bean vine, is found on the edge of monsoon rainforest on basalt slopes and has been recorded from the Kimberley and the Northern Territory.

FIGURE 94   *Backhousia gundarara* is a slender shrub with pink-orange mottled bark and white flowers. An endemic Western Australian species, it is restricted to rainforest surrounding the Mount Jameson massif and is widely disjunct from *Backhousia* species that occur in eastern Australian rainforests. *Image: Russell Barrett*

The Native Grape (*Ampelocissus acetosa*), a rampant vine often found on the edge of rainforest patches, is common across northern Australia and New Guinea.

FIGURE 95

The Native Crepe Myrtle (*Lagerstroemia archeriana*) is a tree or shrub with flowers featuring crinkled petals. Two subspecies have been recorded in Australia: *L. archeriana* var. *divaricatiflora* is found only in the Kimberley monsoon rainforest and *L. archeriana* subsp. *archeriana* in Queensland rainforests extending into New Guinea. This image shows a cultivated specimen.

FIGURE 96

FIGURE 97
(LEFT)

The Arrowroot (*Tacca leontopetaloides*) is a tuberous perennial herb, formerly used as a source of starch in the Pacific area. It has been recorded in Africa, Madagascar, Asia, Pacific Islands and across northern Australia. *Image: Tim Willing*

FIGURE 98
(RIGHT)

The Great Woolly Malayan Lilac (*Callicarpa candicans*) is a small tree with showy purple flowers and fruits that range in colour from purple to black. It has been recorded across most of Asia and northern Australia. *Image: Tim Willing*

Fungi

The rainforest understorey consists of low trees, shrubs and vines. There is virtually no ground flora apart from young seedlings. Leaf litter is seldom more than 2–3 cm thick. Many forms of fungi (such as wood-rotters and mycorrhizal fungi) occur in the rainforest, with fruiting bodies often appearing only in the wet season.

Little is known of fungi in tropical Australia, but there is strong evidence to suggest that Australia has a significant level of endemism in its fungal species.[29] As many fungal species in the Kimberley rainforest are pan-tropical, their identification is extremely difficult. For example, the taxonomy of the genus *Ganoderma* has been described as chaotic, with very few reliable descriptions. However, fungal taxonomy is being transformed by advances in DNA

FIGURE 99 (TOP)

The Hairy Trumpet (*Panus fasciatus*) is a common wood-decaying fungus.

FIGURE 100 (ABOVE)

The bracket fungi *Ganoderma steyaertanum* is one of a group of polypore fungi that grow on wood.

FIGURE 101 (ABOVE RIGHT)

Amanita affin. *hemibapha* is a widespread tropical mycorrhizal fungus that lives on monsoon rainforest trees.

FIGURE 102 (RIGHT)

Boletus (*Tylopilus* species) is a group of bolete mycorrhizal fungi that live on monsoon rainforest trees.

sequencing that will allow researchers to assess the biogeographic and systematic affinities between fungal species.

Because of the difficulty in accessing the Kimberley during the wet season, fungi are rarely collected and studied. Fungi that have been recorded from Kimberley rainforests include the Wood or Pig Ear Fungus (*Auricularia cornea*), *Coprinopsis clastophylla*, *Daldinia concentrica*, the Hairy Trumpet (*Panus fasciatus*), bracket fungi (*Ganoderma steyaertanum*, *Hexagonia tenuis*, *Polyporus arcularius*, *P. tricholoma*, *Trametes muelleri* and *Truncospora ochroleuca*), the tiny *Marasmius* species and *Coltricia albertinii*, all of which grow on dead wood or bark (Figs 99, 100).[30]

Other fungi species that have been collected include *Amanita* affin. *hemibapha* and the *Tylopilus* species of bolete fungi (Figs 101, 102).

Protoxerula flavo-olivacea var. *kimberleyana* is endemic to tropical Australia and was originally described from material collected in deep rainforest litter on North Maret Island.[30] The mushroom genus *Amanita*, which is found on all continents, has a mycorrhizal association with rainforest plants – the fungi are in contact with the plant roots but do not parasitise them. In this symbiotic relationship, the fungi provide the plant with nutrients from the soil, while the plants provide the fungi with photosynthesised sugars.

Bryophytes and lichens

The other non-vascular plants found in the Kimberley rainforest, the bryophytes (which include mosses, liverworts and hornworts), are also difficult to study. Almost all of the specimens collected to date are corticolous – that is, living or growing on the bark of rainforest trees and shrubs. Bryophytes that have been collected from patches on Mitchell Plateau include liverworts (*Riccia multifida*, *Frullania ericoides*, *F. falciloba*, *F. probosciphora* and *F. squarrosula*) and mosses (*Calymperes tenerum*, *Fissidens brassii* var. *hebetatus*, *F. curvatus* var. *curvatus*, *Octoblepharum albidum* and *Trachyphyllum inflexum*).

Corticolous lichens recorded on Mitchell Plateau include *Anisomeridium americanum*, *Collema coccophorum*, *Dirinaria aegialita*, *D. confluens*, *Graphis* species, *Hemithecium implicatum*, *Heterodermia* species, *Lithothelium nanosporum*, *Melanelia fuscosorediata*, *Ramalina subfraxinea*, *Pyrenula nitida* and *Pyxine cocoes*.

Ferns

Ferns are rarely encountered in Kimberley monsoon rainforest and are more typically found in gorges and swamp forest where seepage areas support their growth. Species commonly found in these areas include the Leather Fern (*Acrostichum speciosum*), Oak-leaf Fern (*Drynaria quercifolia*), Swamp Shield Fern (*Cyclosorus interruptus*), Necklace Fern (*Lindsaea ensifolia*), Maidenhair Ferns (*Lygodium flexuosum* and *L. microphyllum*) and Climbing Swamp Fern (*Stenochlaena palustris*) (Figs 103, 104).

Orchids and mistletoes

Only two epiphytic orchids have been recorded in the Kimberley monsoon rainforests. The White Butterfly or Tree Orchid (*Dendrobium affine*) attaches to tree branches in the rainforest canopy (Fig. 105). Endemic to Australia, this species also occurs in the Northern Territory but has not been recorded in Queensland.

The Small-leaf Climbing Fern (*Lygodium microphyllum*) is native to tropical and sub-tropical areas of Africa, South-East Asia and Australia. In some parts of the USA, it is a serious invader of swamps and glades.

FIGURE 103
(TOP LEFT)

The Climbing Swamp Fern (*Stenochlaena palustris*) is native to Australia.

FIGURE 104
(TOP RIGHT)

The White Butterfly Orchid (*Dendrobium affine*), a bark epiphyte that grows on tree branches, is the only true epiphytic orchid recorded thus far from Kimberley monsoon rainforest. It is also found in northern parts of the Northern Territory, including Melville Island, and in Indonesia on the Maluku Islands (Moluccas). It also occurred on Timor, but may now be extinct there.

FIGURE 105
(ABOVE)

FIGURE 106 (ABOVE)

The flowers of the Channel-leaved Tree Orchid (*Cymbidium canaliculatum*) are variably coloured, often with spots and blotches. In the Kimberley, the flowers are mainly brown and cream and are borne on short racemes. The fruit is a green obovoid pod.

FIGURE 107 (LEFT)

The Channel-leaved Tree Orchid, which grows in tree forks and hollows, is endemic to Australia and ranges from the Kimberley across northern Australia and down into New South Wales.

FIGURE 108 (RIGHT)

Nyilinyil or Twin-leaved Mistletoe (*Amyema benthamii*) occurs in Western Australia and the Northern Territory. *Image: Brian Carter*

FIGURE 109 (BOTTOM LEFT)

Yugulu or Love Vine (*Cassytha filiformis*) is a pantropical species that is widespread across northern Australia.

The other epiphytic species is the Channel-leaved Tree Orchid (*Cymbidium canaliculatum*), which occurs on the edges of the coastal monsoon vine thickets of the Dampier Peninsula (Figs 106, 107). It has adapted to survive in a dry habitat, with thick leaves that resist desiccation. The plant has an extensive root and rhizome system that spreads deep into decaying heartwood.[31]

Orchids found on the floor of the Kimberley monsoon rainforest are erect, terrestrial geophytes with underground tubers. They are usually cryptic, have limited geographical ranges and flower during the wet season. They include the Shepherds Crook Orchid (*Eulophia picta*), the Coastal Rein Orchid (*Habenaria hymenophylla*), the Ribbed Shield Orchid (*Nervilia holochila*), the Chinese Spiranthes (*Spiranthes sinensis*) and the Common Jewel Orchid (*Zeuxine oblonga*). The Coastal Rein Orchid is endemic to rainforests across northern Australia, but in the Kimberley it is known only from a single location in the north, a deep gorge near the Prince Regent River. This population is disjunct from the closest known populations in the Northern Territory.[32]

The hemi-parasitic Twin-leaved Mistletoe (*Amyema benthamii*) and the hemi-parasitic *Yugulu* or Love Vine (*Cassytha filiformis*) occur on a variety of host species in the rainforests and monsoon vine thickets throughout the Kimberley (Figs 108, 109).

Native figs

Native figs (*Ficus* species) are one of the most important groups of plants in rainforests for tropical frugivores. Figs possess diverse habits – which describe the shape and appearance of their growth form – many of which are characteristic of tropical rainforest plants such as hemi-epiphytes, a group that includes strangler and banyan figs (*Ficus virens* var. *virens*), large woody climbers and cauliflorous trees (which bear flowers or fruit on their trunks). The ecological advantages of cauliflory are numerous. In many cases, clusters of heavy fruits are produced that would easily be broken off if not attached to strong woody stems. Fruits growing close to the ground are more accessible to animals that cannot climb trees. The tropical naturalist Alfred Russell Wallace argued that cauliflory arose in the dark understorey of the tropical forest as a result of selection for trunk flowers, which are more apparent to pollinators than canopy flowers.[33] Insects on the stem of cauliflorous trees or near the ground can also help pollination.

Figs provide an abundance of food all year and may be critically important to wildlife when other fruits are unavailable. Native figs also have nearly three times more calcium than non-fig fruits.[34]

Beach Berry Bush (*Colubrina asiatica*) is widespread across northern Australia and into New Guinea. *Image: Tim Willing*

FIGURE 110
(LEFT)

Lepisanthes rubiginosa has been recorded in Australia only from Kimberley monsoon rainforests, but it is also found in Asia and Malesia. *Image: Tim Willing*

FIGURE 111
(RIGHT)

Many fig species are pioneer colonisers and have a significant role in establishing rainforest patches.[35]

Rainforest plant distributions

Not every monsoon rainforest patch contains the same suite of species; some species are widespread, while others are restricted or have disjunct distributions. For example, *Mangarr* or Wild Plum (*Sersalisia sericea*) appears to be endemic to Australian rainforests. The tree *Dysoxylum acutangulum* has been recorded in the Kimberley only from Berthier and Maret Islands, although it is common in the Northern Territory and Queensland as well as Timor. Beach Berry Bush (*Colubrina asiatica*), a shrub that is common in rainforest across northern Australia (extending into New Guinea), is known from only three Kimberley rainforest patches, and *Margaritaria dubium-traceyi*, common in rainforests in the Northern Territory and Queensland, has been collected from only one rainforest patch on the Bougainville Peninsula in the Kimberley (Fig. 110). The trees *Lepisanthes rubiginosa* and *Melochia umbellata* occur in South-East Asia and Malesia (an equatorial region including Malaysia, Indonesia, New Guinea, the Philippines and Brunei), but in Australia have been recorded only from monsoon rainforest in the Kimberley (Fig. 111). The trees *Alectryon connatus*, *Euphorbia plumerioides*, Garuga (*Garuga floribunda*), Native Crepe Myrtle (*Lagerstroemia archeriana*) and Climbing Wattle (*Senegalia albizioides*) are common in Kimberley

and north-eastern Australian rainforests but are not found in the Northern Territory.[36,37] Other tree species such as the Australian Jujube (*Ziziphus quadrilocularis*) and *Alectryon kimberleyanus* are endemic Australian species found in the Kimberley and the Northern Territory but not in Queensland. The Ebony (*Diospyros rugulosa*) is a small tree with a similar range in Australia that probably also occurs in Indonesia and Timor. The vine *Ipomoea trichosperma*, a member of the morning glory family (*Convolvulaceae*), is restricted to Kimberley rainforests and Timor. The *Waramburr* or Snake Vine (*Tinospora smilacina*) is endemic to Australian rainforests.

Two species of droopy leaf have been recorded in the Kimberley: *Aglaia brownii* has been found on the Maret Islands, but its range extends across northern Australia and into New Guinea, while *A. eleaeagnoidea* extends from India in the west, throughout much of South-East Asia, and to the Pacific islands of Wallis and Futuna at its eastern-most limit.[38] Molecular studies on droopy leaf species provide evidence of two floristic exchange routes between northern Australia and South-East Asia: one between New Guinea and Cape York Peninsula, and one between Timor-Leste and the Kimberley Plateau. This suggests that the monsoon tropics of the Kimberley Plateau and the monsoon and wet tropics of Cape York Peninsula have separate biogeographic histories.[16]

Six plant species are currently recognised as being endemic to Kimberley rainforest patches, but more remain to be collected and classified.[31,39] Also, the fact that the taxonomy of many pan-tropical plant groups is still unresolved complicates efforts to determine whether a particular species is endemic. The Monsoon Hibiscus (*Hibiscus peralbus*), which is widespread in rainforest patches, belongs to the section of *Hibiscus* known as *Bombicella*, a pan-tropical and subtropical group extending from the Americas through Africa, Asia and Australia with some species extending into the temperate zones (Fig. 112).[40] The small tree *Myrsine kimberleyensis* and the shrub *Midingaran* (*Pavetta kimberleyana*)

are well represented in rainforest patches (Fig. 113). The aromatic shrub *Backhousia gundarara*, found only in one rainforest patch at the southern end of the Mount Jameson massif in the north-west Kimberley, is widely disjunct from the remaining *Backhousia* species that occur in the eastern Australian rainforests. It appears to be a lineage isolated by increasing aridity during the Miocene.[41] Two vines, the Itchy Bean (*Mucuna diabolica* subsp. *kenneallyi*), from Mitchell Plateau, and *Parsonsia kimberleyensis*, from Cape Leveque on the Dampier Peninsula and from some Kimberley islands, are endemic to Kimberley monsoon rainforest.

Another monsoon rainforest plant that occurs in the same area as *B. gundarara* is *Kalanchoe spathulata* var. *spathulata* (Fig. 114). This succulent is known in Australia only from Kimberley rainforest but has a widespread distribution extending through Malesia, Indochina, the Indian subcontinent, Japan and China.[42,43]

FIGURE 115   *Kimboraga exanima* Solem, 1985, a member of the Camaenidae family of land snails, is found in monsoon rainforest on St Andrew Island in St George Basin. *Image: Vince Kessner*

FIGURE 116   *Amplirhagada camdenensis* Köhler, 2010, a member of the Camaenidae family, is endemic to monsoon rainforest on Augustus Island in Camden Sound. *Image: Vince Kessner*

A group of mating snails of the species *Amimopina macleayi* Brazier, 1876, part of the Cerastidae family of land snails, in monsoon rainforest on Bigge Island. *Image: Vince Kessner*

FIGURE 117
(LEFT)

Torresitrachia hartogi Köhler, 2011, from the Camaenidae family, on scree slopes in monsoon rainforest on the islands of the Buccaneer Archipelago. *Image: Vince Kessner*

FIGURE 118
(RIGHT)

RAINFOREST ANIMALS

Invertebrates

Land snails constitute a significant and characteristic element of the world's terrestrial fauna, and they may be the most diverse group after arthropods and nematodes (Figs 115–118). They feature rather seldom in general accounts of biodiversity and its causes, despite the fact that they appear to be among the species most vulnerable to human-induced extinction.[44]

Marine and terrestrial snails were first collected from the Kimberley coast during explorer Phillip Parker King's hydrographic surveys of the Australian coast aboard HMC *Mermaid* between 1819 and 1820. Some of these specimens were deposited in the collections of the British Museum. The first zoological material to be reported from the North Kimberley was collected by entomologist J.J. Walker while he was Chief Engineer aboard the survey ship HMS *Penguin* in 1890 and 1891. Walker collected insects, land snails and other invertebrates – many from Kimberley islands – that were later deposited in the British Museum.[45]

Snail expert Alan Solem was the first person to attempt a global synthesis of land snails.[46,47] Although incomplete and in some cases proven wrong by molecular techniques, Solem's work set an agenda, raising questions of global significance and providing much of the material on which answers could be based. The stimulus for his synthesis came from his intensive studies in Australasia, where his systematic and ecological work prompted questions about the patterns he revealed. Both his basic systematic work and his exploration of the causes of these patterns were incomplete at the time of his death in 1990.[48] His studies in the Kimberley rainforest patches showed that local camaenid faunas usually contained only one species in each genus represented.[49,50]

Almost all the plants, birds and non-camaenid land snails recorded in Kimberley monsoon rainforests are also known from the Northern Territory or Queensland rainforests.[51] They include the majority of canopy trees that protect the rest of the community during the protracted dry season. The biogeographic affinities and lack of endemism in these taxa groups do not suggest prolonged isolation.

However, many of the small organisms are endemic to Kimberley monsoon rainforests, particularly those that are not readily dispersed by other species and that have poor mobility. Examples include the majority of the camaenid land snails,[52] earthworms, certain lizards, and a few small mammals and birds. More recent work on other invertebrate groups indicates that the pattern of small-scale distribution that Solem found for land snails also applies to other groups. For example, Dr Mark Harvey from the Western Australian Museum found 500 species of spiders in the Kimberley, almost all of which had small ranges and were previously undescribed.[53] This kind of information changed the Western Australian Government's emphasis on the importance of short-range species and management of protected areas.[48]

In a study of one Kimberley camaenid snail genus (*Rhagada*), researchers found that the oldest two species lineages (*R. sheai* and *R. worora*) may have originated 3.5 million years ago.[54] The endemic species appear to have survived in Kimberley refugia associated with riverine, mangrove or sandstone habitats. This process of fragmentation created an exceptional diversity for the snails, with similar processes occurring in other groups of animals as well. For instance, the Buff-sided Robin (*Poecilodryas cerviniventris*) and the Rough Brown Rainbow Skink (*Carlia johnstonei*) inhabit monsoon rainforest and riverine vegetation, the Lemon-breasted (Kimberley) Flycatcher (*Microeca flavigaster*) and the rodent *Melomys burtoni* are confined to the mangroves and monsoon rainforest, and the

White-lined Honeyeater (*Meliphaga albilineata*) favours sandstone escarpments that are peripheral to rainforest patches.[1]

Green Tree Ants (*Oecophylla smaragdina*), sometimes called Weaver Ants, are common in the monsoon rainforests of the Kimberley and in Australian rainforests, but not on the Dampier Peninsula (Fig. 119). These aggressive ants build balloon-shaped nests among the foliage of trees and shrubs, which they defend by swarming onto any unfortunate intruder and attacking with great ferocity. They cannot sting, but bite with their jaws and squirt a burning fluid onto the wound. Aboriginal people collect the nests and crush them in water to make a thirst-quenching drink that has an astringent, but not unpleasant, taste.

Research on the ant fauna of the monsoon vine thickets of the Dampier Peninsula has shown that frequent fire promotes structural changes to vegetation, including a more open canopy, which favour arid-adapted ant taxa.[55] By contrast, long periods without fire encourage more shade-tolerant, forest-associated species. As the ecosystem changes over time to one featuring more rainforest, a different ant fauna develops.

A balloon-shaped nest of *Yuulway, Yulwe* or Green Tree Ants (*Oecophylla smaragdina*). *Image: Tim Willing*   FIGURE 119

MONSOON RAINFOREST ONLY	Australian Figbird	Orange-footed Scrubfowl
	Common Cicadabird	Rainbow Pitta
	Emerald Dove	Rufous Owl
	Green Oriole	Spangled Drongo
	Green-backed Gerygone	Varied Triller
	Little Shrike-thrush	
MONSOON RAINFOREST AND RIVERINE FOREST	Brush Cuckoo	Common Koel
	Buff-sided Robin	Oriental Cuckoo
MONSOON RAINFOREST AND MANGROVES	Broad-billed Flycatcher	Rose-crowned Fruit-dove
	Little Bronze Cuckoo	Shining Flycatcher
	Mangrove Golden Whistler	Wood Fantail
	Pied Imperial Pigeon	

TABLE 1   Birds that rely on Kimberley monsoon rainforests or closed-canopy communities

During the dry season, insect activity is suppressed by a combination of factors: low atmospheric and forest floor humidity; the drying up of freshwater streams and pools; diminished plant growth; scarcity of flowers and fruits; and cooler temperatures (especially at night in inland areas). Many insects pass this relatively inhospitable season as dormant immature stages concealed in the soil, in nests or within plant tissue.[56]

Birds

Many birds rely on the Kimberley monsoon rainforests (Table 1). These habitats are important for the seasonal migration of birds from Indonesia, such as the Common Koel (*Eudynamys scolopacea*), the Dollarbird (*Eurystomus orientalis*) and the Channel-billed Cuckoo (*Scythrops novaehollandiae*). Frugivorous birds such as the Rose-

The *Mandamanda* or Rose-crowned Fruit-dove (*Ptilinopus regina*) is widespread across northern and eastern Australia and is also found in Indonesia. *Image: Lochman Transparencies*

FIGURE 120

crowned Fruit-dove and Emerald Dove (*Chalcophaps indica*) are fairly sedentary, whereas the Pied Imperial Pigeon is nomadic in the Kimberley and widespread in northern and eastern Australia, moving and feeding among patches of monsoonal rainforest (Figs 120-122). Frugivorous birds require many patches to maintain their populations.[57]

Other birds forage high in the canopy and depend on the fleshy fruits of the rainforest. The Australasian Figbird (*Sphecotheres vieilloti*) eats mostly figs, as its name suggests, but also a wide variety of other fruits. It has a wide distribution across northern and eastern Australia, southern New Guinea, and the Kai Islands of Indonesia. The Common Cicadabird (*Edolisoma tenuirostre*) is found in northern Australia, Indonesia, New Guinea and the Solomon Islands. The Green Oriole or Australian Yellow Oriole (*Oriolus flavocinctus*) is an inconspicuous inhabitant of lush tropical vegetation throughout Australia and New Guinea. Green Orioles forage slowly and methodically through the middle and upper strata of dense forests, feeding mainly on fruit. On Mitchell Plateau, they have been observed eating Green Tree Ants. The Green-backed Gerygone (*Gerygone chloronota*) is found in northern Australia and New Guinea. Its natural habitats are subtropical or tropical moist lowland rainforests, dense riverine forest, and subtropical or tropical mangrove forests; on Mitchell Plateau, this species favours the edges of rainforest and thickets fringing watercourses. The Little Shrike-thrush or Arafura Shrike Thrush (*Colluricincla megarhyncha*) is found in the top end of the Northern Territory and in the north-west Kimberley in rainforest, vine thickets and mangroves. It has been recorded eating molluscs and berries. The Spangled Drongo (*Dicrurus bracteatus*) is found throughout northern and eastern Australia and in New Guinea and eastern Indonesia. Individuals from the northern areas of Western Australia and the Northern Territory migrate northwards to Indonesia. Spangled Drongos prefer wet forests but can also be found in other woodlands and mangroves. They feed on insects, fruit and nectar. The Varied Triller (*Lalage leucomela*) feeds mainly on fruit, mostly in the outer foliage of trees but occasionally on the ground or on tree trunks near the ground. It is found in New Guinea and Australia, where its distribution stretches from the north-west Kimberley to Cape York Peninsula, and down the east coast to around Sydney.

Some birds, such as the Rufous Boobook or Rufous Owl (*Ninox rufa*), use the rainforest canopy to roost during the day and hunt at night (Fig. 123). Named after the rust-coloured feathers of mature birds, the Rufous Owl is the only exclusively tropical owl in Australia. It is a shy and elusive bird that generally roosts alone or in pairs in thickly foliaged mature trees with a good view of the surroundings. A skilled and powerful hunter, the Rufous Owl can capture a wide variety of prey, ranging from birds and insects to small mammals such as flying foxes. Although the species is uncommon, it has a wide geographic range. It is native to the Indonesian Aru Islands, New Guinea and northern Australia, where it is found in Arnhem Land, the North Kimberley, the eastern Cape York Peninsula and the Mackay District of eastern Queensland.

Other birds feed on the rainforest floor. The Emerald Dove is quite terrestrial, often searching for fallen fruit and seeds on the ground and spending little time in trees (except when roosting). It is easily identified by its shiny green wings and white patch on each shoulder. The Rainbow Pitta (*Pitta iris*) forages on the forest floor for insects and their larvae, other arthropods, snails and earthworms (Fig. 124).

FIGURE 121 (TOP)

The Emerald Dove (*Chalcophaps indica*) is usually found alone, in a pair or in small groups. Emerald Doves move between the monsoon rainforests of Indonesia and the Kimberley. *Image: Lochman Transparencies*

FIGURE 122 (ABOVE)

The *Nyulbu* or Pied Imperial Pigeon (*Ducula bicolor*) is found in South-East Asia, the Philippines, Indonesia, New Guinea and northern Australia – from the Kimberley to the Cape York Peninsula, and along the east coast from the Torres Strait to Central Queensland. *Image: Ron Johnstone, WA Museum*

FIGURE 123 (ABOVE)

The Rufous Owl or Rufous Boobook (*Ninox rufa*) is Australia's second largest owl and the largest in the tropical north. It is a powerful night hunter of arboreal mammals such as possums, Sugar Gliders and flying foxes. Rufous Owls are found alone or in pairs in mature trees with thick foliage that offer a good view of the surroundings. *Image: Lochman Transparencies*

FIGURE 124   The Rainbow Pitta (*Pitta iris johnstoneiana*) is a secretive and shy bird that feeds mainly on insects and small vertebrates. The species is endemic to northern Australia. *Image: Lochman Transparencies*

FIGURE 125   The Orange-footed Scrubfowl (*Megapodius reinwardt*) feeds on seeds, fallen fruit and terrestrial invertebrates. The species is found across northern Australia, New Guinea, Indonesia and the Philippines. *Image: Lochman Transparencies*

The species is endemic to northern Australia and most closely related to the Superb Pitta of Manus Island in northern Papua New Guinea. In Western Australia, the Rainbow Pitta is restricted to dense rainforest patches along the North Kimberley from Walcott Inlet to Napier Broome Bay. It is also found on some of the islands of the Bonaparte Archipelago. The Orange-footed Scrubfowl (*Megapodius reinwardt*) is a terrestrial, mound-building bird that is common in suburban Darwin gardens – where it is also called a Bush Chook or Bush Turkey (Fig. 125). It uses a range of forest and scrub habitats, where it feeds on seeds, fallen fruit and terrestrial invertebrates. It spends the day foraging on the forest floor and will only fly if disturbed or to roost in trees. As with other megapodes, the Orange-footed Scrubfowl builds a large nest – up to 8.5 metres across at its base – using large mounds of sand, leaf litter and other debris. The heat generated by the decomposition of this organic material serves to incubate the bird's eggs.

Monsoon rainforest on sand dunes and scree slopes below a laterite plateau on North Maret Island.

CHAPTER 8
Rainforest on sand dunes

The monsoon rainforest on the coastal Holocene sand dunes of the Dampier Peninsula is listed as a vulnerable ecological community on the Western Australian Government's list of Threatened Ecological Communities, along with the rainforest swamps at Theda, Walcott Inlet and the Roe River. In 2013, the Dampier Peninsula monsoon rainforest was recognised by the Australian Government as Nationally Endangered under Section 184 of the *Environment Protection and Biodiversity Conservation Act 1999*.

Directly within and behind the swales of the coastal Holocene dune system north of Broome, on both the west and east coasts of the Dampier Peninsula, 79 discontinuous but discrete pockets of monsoon rainforest occur as long narrow patches with exposed edges.[1] The patches vary in size: although most patches do not exceed 0.1 square kilometres, one patch is 0.9 square kilometres and five patches are larger than 1 square kilometre. The largest patches occur at the far northern end of the Peninsula above the 750 mm per annum rainfall zone and include the most floristically species-rich patches. The total area of the 79 patches is less than 2.6 square kilometres. Although they represent less than 0.001% of the total land area of the Dampier Peninsula, the patches contain almost one-quarter of the total plant species recorded for the Peninsula (Figs 126, 127).

FIGURE 126

Floodwaters pooling behind the coastal sand dunes at Broome's Minyirr Park, an area that supports patches of monsoon rainforest. Broome had its wettest January on record in 2022, with 941 mm of monsoon and cyclonic rainfall.

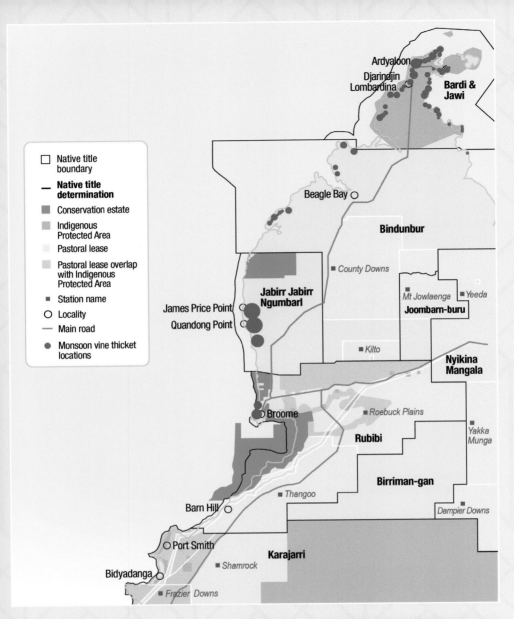

Legend:

- ☐ Native title boundary
- ▬ **Native title determination**
- ■ Conservation estate
- ▨ Indigenous Protected Area
- ▢ Pastoral lease
- ▨ Pastoral lease overlap with Indigenous Protected Area
- ■ Station name
- ○ Locality
- — Main road
- ● Monsoon vine thicket locations

Ardyaloon
Djarindjin
Lombardina
Bardi & Jawi

Beagle Bay

Bindunbur

■ County Downs

■ Mt Jowlaenga   ■ Yeeda

Jabirr Jabirr Ngumbarl

Joombarn-buru

James Price Point
Quandong Point

■ Kilto

Nyikina Mangala

■ Roebuck Plains

Broome

■ Yakka Munga

Rubibi

Birriman-gan

■ Thangoo

Barn Hill

■ Dampier Downs

○ Port Smith

Karajarri

■ Shamrock

Bidyadanga

■ Frazier Downs

FIGURE 127 (ABOVE)

Map of the Dampier Peninsula showing the distribution of monsoon forest on or behind coastal sand dunes. *Image: Environs Kimberley*

FIGURE 128 (RIGHT)

Cable Beach Ghost Gum (*Corymbia paractia*) is an endemic species that is restricted to the Cable Beach area in Broome. *Image: Tim Willing*

FIGURE 129 (ABOVE)

Tuckeroo (*Cupaniopsis anacardioides*) is endemic to northern Australia.

FIGURE 130 (RIGHT AND ABOVE RIGHT)

In the Kimberley, *Pittosporum moluccanum* is restricted to coastal beach monsoon rainforests. It is also found in the Northern Territory, North-East Queensland and Asia.

These patches are allied to monsoon rainforest but are floristically different from the rainforest patches in the wetter north-west Kimberley. They range from semi-deciduous monsoon vine thicket to closed semi-deciduous monsoon rainforest.[2] Patches around Broome also include the endemic Cable Beach Ghost Gum (*Corymbia paractia*) (Fig. 128). North of Weedong Well, on the west coast of the Dampier Peninsula, trees species can include *Goonj* (*Celtis strychnoides*), *Croton habrophyllus*, *Jarnba* or Mistletoe Tree (*Exocarpos latifolius*), *Garnboorr* (*Melaleuca dealbata*), *Marool* or Blackberry Tree (*Terminalia petiolaris*) and *Mamajen* (*Mimusops elengi*). In these coastal thickets, vines are more evident from inside the patch rather than in the canopy and include *Capparis lasiantha*, *Gymnanthera oblonga*, the Kimberley endemic *Parsonsia kimberleyensis*, Stinking Passionflower (*Passiflora foetida*), *Waramburr* or Snake Vine (*Tinospora smilacina*) and *Secamone elliptica*.

The monsoon vine thicket communities are best developed at the northern end of the Dampier Peninsula, especially along the western coastline. Emergent trees are up to 12 metres tall and are principally the *Marool* or Blackberry Tree, *Mangarr* and, on the landward fringe, the *Gubinge* (*Terminalia ferdinandiana*). A hybrid between the *Gubinge* and the *Marool* is known to the local Aboriginal people as Red Gubinge and is highly valued, as the fruit is much fuller and sweeter than that of its parent trees.

DAMPIER PENINSULA - MONSOON
RAINFOREST VEGETATION TRANSECT

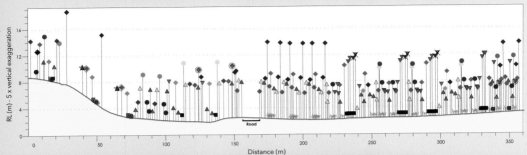

Grasses

- Cenchrus ciliaris
- Cymbopogon sp.
- Enneapogon pallidus
- Enneapogon sp.
- Enteropogon dolichostachyus
- Eriachne sp. nov.

Medium to Large Shrubs

- ▼ Acacia colei
- ▼ Acacia eriopoda
- ▼ Acacia monticola
- ▲ Breynia cernua
- △ Bridelia tomentosa
- △ Caesalpinia major
- △ Calytrix exstipulata
- ▲ Capparis lasiantha
- ▲ Distichostemon hispidulus var. aridus
- ▲ Flueggea virosa subsp. melanthesoides
- ▲ Glycosmis sp.
- ▲ Sersalisia sericea

Small Shrubs

- ■ Abutilon indicum var. australiense
- ■ Achyranthes aspera
- ■ Boerhavia sp.
- ■ Cleome viscosa
- ■ Corchorus sidoides subsp. vermicularis
- ■ Crotalaria sp.
- ■ Gomphrena pusilla
- ■ Hypoestes floribunda var. varia
- ■ Indigofera colutea
- ■ Indigofera linnaei
- ■ Jasminum molle
- ■ Josephinia eugeniae
- ■ Microstachys chamaelea
- ■ Sida hackettiana
- ■ Tephrosia rosea var. rosea

Trees

- ◆ Bauhinia cunninghamii
- ◆ Celtis philippensis
- ◆ Codonocarpus cotinifolius
- ◆ Corymbia bella
- ◆ Corymbia greeniana
- ◆ Croton habrophyllus
- ◆ Diospiros humilis
- ◆ Ehretia saligna var. saligna
- ◆ Exocarpos latifolius
- ◆ Grewia breviflora
- Gyrocarpus americanus subsp. pachyphyllus
- ● Hakea macrocarpa
- ● Mimusops elengi
- ● Pavetta kimberleyana
- ● Premna acuminata
- ● Santalum lanceolatum
- ● Terminalia ferdinandiana
- ● Terminalia petiolaris

Climbers

- ◇ Abrus precatorius subsp. precatorius
- ⬡ Amyema benthamii
- ☆ Capparis lasiantha
- ◎ Cassytha sp.
- ✪ Jacquemontia paniculata
- ⊠ Leichhardtia viridiflora
- △ Passiflora foetida var. hispida
- ✛ Tinospera smilacina
- ⊕ Tylophora cinerascens

- ✕ Dead

Other trees in the monsoon vine thicket communities along the Dampier Peninsula coast include the Banyan Fig (*Ficus virens* var. *virens*), *Ficus geniculata* var. *insignis* and the Wild Apple (*Syzygium eucalyptoides* subsp. *bleeseri*). Species confined to localised populations include the Tuckeroo (*Cupaniopsis anacardioides*), Broad Leaved Ebony (*Diospyros maritima*), *D. rugulosa*, *Pittosporum moluccanum*, *Trophis scandens* and *Vitex glabrata* (Figs 129, 130).

Fruiting times vary between patches of monsoon vine thicket along the coast, which are too small to support a large number of animal species. Birds, bats, and the *Boorroo* or Agile Wallaby (*Macropus agilis*) move between patches, eating fruits and spreading seeds, thus keeping the patches ecologically connected. The endangered Gouldian Finch (*Chloebia gouldiae*) occurs on the margins of the patches. The Dampierland Burrowing Snake (*Simoselaps minimus*) and the Dampierland Limbless Slider (*Lerista apoda*) are both Kimberley endemic species found only within and adjacent to monsoon vine thicket patches on the Dampier Peninsula.

Pindan, the ubiquitous vegetation that dominates the red sandplains of the Dampier Peninsula, often intrudes into or separates patches of coastal monsoon thickets (Fig. 131). It is grassland covered by a sparse upper layer composed mainly of eucalypts, with a dense, often thicket-forming, middle layer of predominantly wattles. It is in the fringing pindan that the much sought-after edible seed pods of *Magabala*, *Garlarla* or Bush Banana (*Leichhardtia viridiflora* subsp. *tropica*) are found.

Small patches of monsoon rainforest have also been recorded in coastal dunes on North Maret Island (and adjacent Berthier Island) off the Kimberley coast. *Pittosporum moluccanum*, a plant species known from the Dampier Peninsula and restricted to coastal dune monsoon forest, has been recorded in both locations. Small patches also occur on Holocene marine dunes on the seaward edge of tidal mudflats that fringe the Timor Sea, east of Cape Domett on the far north coast. Coastal sand dunes are a very common habitat for monsoon forest in tropical Australia; its near absence from the north-west Kimberley is most likely due to the rarity of coastal dunes.

FIGURE 131 (OPPOSITE)

The diversity of vegetation in monsoon rainforest is visible in this transect north of Broome. On the left of the road is coastal monsoon rainforest vine thicket, and on the right is pindan Acacia shrubland. *Source: Biota Environmental*

MAPPING RAINFOREST PATCHES ON COASTAL DUNES

The importance to science of rainforest patches on coastal dunes was first recognised during a biological survey of the Dampier Peninsula in 1977.[2] Further intensive field studies were undertaken by Kevin Kenneally, assisted by Brian Carter, Dave Dureau, Paul Foulkes and Tim Willing (all members of the Broome Botanical Society), during the preparation of the book *Broome and beyond*.[3] As part of their research, the team worked with Aboriginal communities living on the Dampier Peninsula to understand their relationship with the coastal rainforest patches. Meetings were held with Sandy and Esther Paddy, the Broome Community, Lombadina Community, Beagle Bay Community and Bardi Community, who all supported the project with information on plants, their uses and names (Figs 132, 133). Brian and Violet Carter at Ardyaloon (One Arm Point) facilitated the team's studies on Bardi Country and introduced them to traditional carving of implements from native woods by residents Peter Hunter, Ashley Hunter and Frank Davey (Figs 134, 135). Discussions were also held with lawman Paddy Roe in Broome to understand the cultural connections between the Lurujarri Heritage Trail and the coastal rainforest patches (Fig. 136).

Rainforest patches on the coastal dunes were further investigated during the Kimberley rainforest survey undertaken between June 1987 and March 1989 as part of the National Rainforest Program, funded by the Commonwealth Government and coordinated by the Department of Conservation and Land Management.[4] From 2000 to 2002,

Broome Botanical Society members Sally Black, Tim Willing and Dave Dureau identified and mapped these patches.[5] Detailed mapping of patches was also undertaken by Environs Kimberley in collaboration with Indigenous rangers, communities and Traditional Owners as part of the Kimberley Nature Project.[1]

In 1992, a vine thicket walking trail was opened at Gubinge Road near Cable Beach in Broome. Funded by the National Rainforest Conservation Program, it was designed to enhance public understanding of these vine thickets, which are the most southerly and accessible in the Kimberley.

FIGURE 132 (OPPOSITE, TOP)

Paul Foulkes (1945-1998) was a founding member of the Broome Botanical Society, pioneered Broome bushwalks including the monsoon rainforests and dinosaur trackways, and illustrated plants for the *Broome Advertiser*.

FIGURE 133 (OPPOSITE, MIDDLE)

Brian Carter (1931-2022), Tim Willing and Dave Dureau (1934-2019) were founding members of the Broome Botanical Society and contributed to the understanding of the Dampier Peninsula monsoon rainforest through their tireless field collecting and observations.

FIGURE 134 (OPPOSITE, BOTTOM)

Violet Carter, a Bardi woman from Ardyaloon (One Arm Point), an Aboriginal community on the northern tip of the Dampier Peninsula, home of the Bardi and Jawi people. Violet and her husband Brian provided guidance and support on Country for many years and shared insights into the local culture, passing on a wealth of information on local uses of plants and animals and their Indigenous names.

FIGURE 135 (TOP)

Peter Hunter carving a boomerang from *Hakea arborescens* at Ardyaloon (One Arm Point) in 1992. Shields were made from *Gyrocarpus* and *Canarium*. *Image: Brian Carter*

FIGURE 136 (ABOVE)

Paddy Roe (1912-2001) was a Nyikina man who established the Lurujarri Heritage Trail in 1987 as a way of sharing the cultural importance of the coastal dunes landscape with non-Aboriginal people. *Image: Richard Woldendorp, courtesy of the WA Museum*

Lone Dingo rainforest patch on Mitchell Plateau.

CHAPTER 9
Mitchell Plateau: A focus area

Mitchell Plateau contains the most diverse landforms and habitats in north-western Australia. It also has the highest diversity of vertebrate fauna recorded for any comparable area in Western Australia. Its open forests, particularly on the Cainozoic laterite plateau surfaces, are dominated by the fan palm *Dangana* (*Livistona eastonii*) and have the richest and most diverse mammal assemblage. In contrast, its monsoon rainforest patches and sandstone contain relatively stable populations of fewer species of small mammal, and several habitats had seasonally variable populations and species.[1,2] The biological surveys undertaken on and around Mitchell Plateau from June 1981 to December 1982 highlighted this diversity of species.[1] Intensive ecological studies that began in the early 1980s in areas including Mitchell Plateau, Anjo Peninsula, Purnululu and the Dampier Peninsula collected a range of plant and animals species that clearly showed the advantages of an intensive seasonal and habitat-focused approach to sampling.[3]

There is growing evidence that several vertebrate groups in tropical Australia have undergone marked population declines.[4] In the Kimberley, changes in land use and other practices over the past century have coincided with a wave of local extinctions of medium-sized mammals as well as reductions in the abundance of some small mammal and bird species.[5] These declines are variously postulated as arising from climate change and decreasing groundwater levels; from habitat alteration due to a combination of pastoral grazing by large herbivores (as well as feral cattle, donkeys and pigs) and changed fire regimes (particularly the removal of the shrub layer of tropical savannas); or from predation by feral cats.[6]

The lack of rainforest-specialist fauna in north-western Australia is believed to be due to the lack of large tracts (>10 square kilometres) of monsoon rainforest habitat, the possible substantial contraction of these habitats in the past, and the limited extent of well-developed gallery forest. No mammal species has been recorded

FIGURE 137
(TOP LEFT)
The Rough Brown Rainbow Skink or Johnstone's Carlia (*Carlia johnstonei*) is a Kimberley endemic that is very common in rainforest patches. *Image: Norm McKenzie*

FIGURE 138
(TOP RIGHT)
The Rough-scaled Python (*Morelia carinata*) is endemic to the rainforest patches of the Kimberley. *Image: Ron Johnstone, WA Museum*

FIGURE 139
(ABOVE)
The Giant Slender Blue-tongue (*Cyclodomorphus maximus*) is found in rainforest patches in the far north Kimberley. *Image: Ron Johnstone, WA Museum*

as being confined to Kimberley monsoon rainforests, which is not surprising considering their relatively small total area. However, these rainforests are used by eutherian habitat generalists (murids and bats) that most commonly occur in surrounding savanna habitats. The mammal assemblages in monsoon rainforests across northern Australia (including Cape York Peninsula, the Northern Territory and the Kimberley) are essentially regional subsets of the local savanna and mangrove mammal assemblages, and consequently have only a limited number of species in common, most of which are bats.[7]

The Kimberley has a rich herpetofauna, with approximately 42 reptiles and about half that number of frogs endemic to the region. However,

FIGURE 140 (TOP)

The Carpet Python (*Morelia spilota variegata*) is common in rainforest across northern Australia and in New Guinea. *Image: Ron Johnstone, WA Museum*

FIGURE 141 (TOP RIGHT)

Vince Kessner, a research associate at the Australian Museum, holding an Olive Python (*Liasis olivaceus*) that had just killed a Mertens' Water Monitor outside a rainforest patch in the East Kimberley. Some hours later, only the tip of the monitor's tail protruded from the snake's mouth.

FIGURE 142 (ABOVE)

The Brown Tree Snake (*Boiga irregularis*) is common in rainforest across northern Australia, New Guinea and Indonesia. *Image: Norm McKenzie*

FIGURE 143 (ABOVE)

The Mertens' Water Monitor (*Varanus mertensi*) is endemic to northern Australia and is always found near fresh water. *Image: Tim Willing*

no herpetofauna species are entirely restricted to monsoon rainforest habitat. Only one species, the Rough Brown Rainbow Skink (*Carlia johnstonei*), seems to occur at higher density in rainforest patches than in other habitats (Fig. 137).[8] Other species recorded in rainforest include the Rough-scaled Python (*Morelia carinata*) and Giant Slender Blue-tongue (*Cyclodomorphus maximus*), both of which are endemic to the Kimberley (Figs 138, 139). The Carpet Python (*Morelia spilota variegata*), Olive Python (*Liasis olivaceus*) and Brown Tree Snake (*Boiga irregularis*) have a more widespread distribution in the tropics but are commonly encountered in rainforest and riparian patches (Figs 140-143).

FIGURE 144
(TOP LEFT)

The rodent *Munjul* or Grassland Melomys (*Melomys burtoni*) is found in Papua New Guinea and Australia, along the coast from the Kimberley to New South Wales. *Image: Norm McKenzie*

FIGURE 145
(TOP RIGHT)

The *Gamurnda*, *Gamundee*, *Langguman* or Savanna Glider (*Petaurus ariel*) is a small omnivorous, arboreal and nocturnal possum. Its name refers to its attraction to sugary foods, such as sap and nectar, and its ability to glide through the air. It is found in New Guinea and in northern and eastern Australia. *Image: Norm McKenzie*

FIGURE 146
(ABOVE)

Once found throughout northern Australia, the *Bundilarri* or Golden Bandicoot (*Isoodon auratus*) is now restricted to small areas of the Northern Territory and Western Australia, including the Kimberley. Populations of these small, rat-like marsupials have faced threats from feral cats and changes in fire regimes. *Image: Norm McKenzie*

FIGURE 147
(BELOW)

The Little Northern Native Cat or Northern Quoll (*Dasyurus hallucatus*) is considered carnivorous, although it feeds primarily on insects. Its diet also includes other small mammals, birds, frogs, reptiles and sometimes fleshy fruits. Northern Quolls are most often found in rocky escarpments and open eucalyptus forests of lowland savannas, but they have been recorded in monsoon rainforest on Mitchell Plateau. *Image: Lochman Transparencies*

FIGURE 148 (ABOVE)

The *Wunggangbarn*, *Jari* or Golden-backed Tree Rat (*Mesembriomys macrurus*) was once widespread across northern Australia but is now restricted to the Kimberley coast and nearly islands. The rodent is largely nocturnal and feeds on seeds, fruits, leaves, grasses and invertebrates. *Image: Norm McKenzie*

FIGURE 149 (TOP)

The *Yilangal* or Scaly-tailed Possum (*Wyulda squamicaudata*) is endemic to the Kimberley and lives on rocky terrain with dense thickets of vines and fruiting trees. The possum uses its tail, which is covered with rasp-like scales, to grasp on to tree branches as it forages for food. *Image: Lochman Transparencies*

FIGURE 150 (ABOVE)

The *Niimanboorr*, *Jarringuu*, *Miniwarra* or Black Flying Fox (*Pteropus alecto*) is native to Australia, Papua New Guinea and Indonesia. During the day, hundreds of thousands of individuals rest in communal roosts, known as camps. At sunset, they fly off to feed on nectar, fruit and blossoms.

FIGURE 151 (TOP)

The Northern Blossom Bat (*Macroglossus minimus*) has a wide geographical range that includes Thailand, Malaysia, southern Philippines, Java, Borneo, New Guinea, the Soloman Islands and northern Australia.

FIGURE 152 (ABOVE)

The Yellow-lipped Cave Bat (*Vespedelus douglasorum*) has been recorded only in the Kimberley, where it roosts in sandstone and limestone caves. *Image: Norm McKenzie*

FIGURE 153 (ABOVE)

The Northern Brushtail Possum (*Trichosurus vulpecula arnhemensis*) lives in mostly forested habitat across northern Australia, including a small number of sites in the Kimberley. On Mitchell Plateau, it lives in eucalypt forest and woodlands and on the edge of monsoon rainforest patches. The possum feeds on a variety of plants, leaves, flowers and fruits, and sometimes even nesting birds. *Image: Lochman Transparencies*

However, at Mitchell Plateau, the endemic Kimberley Rock Rat (*Zyzomys woodwardi*) and the Grassland Melomys (*Melomys burtoni*) have frequently been trapped in monsoon rainforests (Fig. 144). These patches, with their relatively closed canopy and shelter, are used by a wide range of mammals, including the Sugar Glider (*Petaurus breviceps*), Northern Brown Bandicoot (*Isoodon macrourus*), Little Northern Native Cat or Northern Quoll (*Dasyurus hallucatus*), Tunney's Rat (*Rattus tunneyi*), Golden-backed Tree Rat (*Mesembriomys macrurus*) and Common Rock Rat (*Zyzomys argurus*) (Figs 145-148). The *Yilangal* or Scaly-tailed Possum (*Wyulda squamicaudata*), also a Kimberley endemic, is nocturnal and feeds on leaves, flowers and fruits (Fig. 149). It occurs in areas of rugged sandstone and feeds in adjacent woodland or closed forest as well as in patches in sandstone gorges that support monsoon rainforest plant species. Bats such as the Black Flying Fox (*Pteropus alecto*), Little Red Flying-Fox (*Pteropus scapulatus*), Northern Blossom-Bat (*Macroglossus minimus*) and the Yellow-lipped Cave Bat (*Vespadelus douglasorum*) have all been recorded feeding in rainforest patches (Figs 150-152). The Grassland Melomys and the Northern Brushtail Possum (*Trichosurus arnhemensis*), both arboreal feeders, have been observed at night in the seaward fringe of mangroves at Walsh Point on Mitchell Plateau (Fig. 153).[1]

Unnamed members of the 1921 Easton surveying expedition to the Kimberley. *Image: Bill Easton*

CHAPTER 10
Ethical and sustainable use of the Kimberley's biological resources

The plants of the Kimberley have been used for millennia as a source of food and medicine, and more recently have been commercially exploited for their potential pharmaceutical properties.

THE LARDER IN THE LANDSCAPE

Around the world, environmental conservation directives are mandating greater inclusion of Indigenous peoples and their knowledge in the management of global ecosystems.[1] Traditional bush foods and medicinal plants (Indigenous Biocultural Knowledge) play an integral role in Aboriginal cultures. This knowledge is not static but continues to evolve through active social and economic interaction between people and plants in the landscape. Aboriginal peoples carry out the traditional practice of 'wild harvesting' – gathering plants that have not been subjected to domestication or cultivation. Such gathering is aligned with the optimum accumulation of valuable nutritive and biologically active substances in plants. Their harvesting is often from the same locality and is dependent on seasonal availability as well as an acute awareness and management of where in the landscape different plants grow. The locales in which the harvested plants naturally occur are defined as an 'ecoscape', and the practice of wild harvesting is referred to as 'ecoscaping' (Fig. 154).[2]

For Aboriginal peoples, rainforests are important sources of seasonal fruits, berries and yams, as well as timber for carving and traditional medicines.[3] In the Broome and Dampier Peninsula region, Nyul Yul,

FIGURE 154   Although *Gubinge* (*Terminalia ferdinandiana*) may be collected in Yawuru Conservation Park, its sale is prohibited. *Image: Tim Willing*

Yawuru and Bardi Jawi Oorany women ranger groups are being taught to collect, store and propagate culturally significant seeds and endangered plants to help protect the biodiversity.[4] This project is bringing western science and traditional knowledge together.[5] As Bardi Jawi Oorany ranger coordinator Debbie Sibosado said: "It's really important so we learn to know what to do in the Kartiya (white-fella) way, learn the common names, scientific names, so that when we're speaking to people, we all know what we're talking about."[4] After decades of indifference, non-Indigenous Australians are now looking more favourably on native rainforest plants as food sources.[6] Species gaining in popularity include Finger Lime (*Citrus australasica*), *Gubinge*, *Macadamia* species and the Peanut Tree (*Sterculia quadrifida*).

In response to concerns about increasingly popular bush foods, which in Australia are regulated as foods rather than as medicine, the Rural Industries Research and Development Corporation reviewed the safety of Australian bush foods.[7] Now known as AgriFutures Australia, the corporation critically examined botanical, chemical and toxicological information on the major bush foods as well as their history of use in Aboriginal cultures. Its analyses generally confirmed the safety of the bush foods tested, with many adverse effects attributable to the consumption of foods that were unripe or overly acidic, poorly prepared or mistakenly identified.

Aboriginal communities have also helped other researchers to investigate Indigenous medicinal knowledge. For example, Western Sydney University researchers collaborated with the Mbabaram community in far north Queensland and the Yirralka Rangers of North-East Arnhem Land in the Northern Territory to understand the antioxidant and antimicrobial characteristics of eight plant species.[8] Their work identified the need for clearer guidelines and regulations around community-driven biomedical research and established a framework for future collaborations between academic researchers and Indigenous groups.

BIOPROSPECTING AND BIOPIRACY

The purpose of conserving plants and their genetic material – in both the ecosystem and dedicated seed banks – is to ensure their use now and into the future.[9] Some rainforest plants are a rich source of medicinal chemicals such as alkaloids, amides, flavonoids, lignans, sterols and terpenes. The genus *Zanthoxylum*, represented in the Kimberley by the Indian Prickly Ash (*Z. rhetsa*) has been described as "a stockpile of biological and ethnomedicinal properties" and has been commonly used traditionally outside Australia in various ethnomedicines for different ailments.[10] The Dutchman's Pipe (*Aristolochia acuminata*), a vigorous scrambling vine, is common in the rainforests across northern Australia and extending north into the tropics. Although it is used widely in South-East Asia, the Philippines and China for medicinal purposes, its supply, sale or use in therapeutic goods is prohibited in Australia. *Aristolochia* species and particular components called aristolochic acids have been linked to severe kidney damage and urinary tract cancer.[11] Another group of climbing vines found both within and on the edge of the rainforest, *Mucuna*, is regarded as an unconventional plant genus, because its bioactive constituents have a wide range of potentially promising nutritional, pharmaceutical and cosmeceutical applications.[12]

However, Australian bush foods have been subject to claims of biopiracy – when people use a biological resource or traditional knowledge that is linked to a Traditional group, without seeking the group's consent or sharing the resulting benefits with them. Cosmetics company Mary Kay Inc., which is based in the USA but has an office in Melbourne, filed an Australian patent application in 2007 covering topical skincare compositions that use extracts from Kakadu Plum (*Gubinge*). The plant has long been used by the

Mirrar people in Kakadu and elsewhere in the Northern Territory for food and medicine. The company defended its use of the plum, insisting it followed the correct procedures for obtaining extracts.[13] Although Mary Kay Inc. had not contravened any legislation, inconsistent Commonwealth laws meant the company was able to remove plum samples from Australia without negotiating with Indigenous communities.[14] Professor Daniel Robinson, a researcher at the University of New South Wales who focuses on the regulation of nature and knowledge, filed a formal challenge to the patent application with IP Australia, citing evidence of prior use and an established market for products using Kakadu Plum.[13] Mary Kay Inc. withdrew its Australian patent application in 2011, but a similar US patent was granted in 2020.[15,16]

The Nagoya Protocol to the Convention on Biological Diversity requires companies that use flora or fauna in their products to acknowledge their origin and share the profits with any Indigenous peoples and local communities whose knowledge made the development of these products possible.[17] However, as the USA is not a signatory to this international agreement, US companies such as Mary Kay Inc. are not subject to the obligations created by the Protocol.

The *Biodiversity Conservation Bill 2015* was introduced to the Western Australian Parliament by the Minister for the Environment to repeal the *Wildlife Conservation Act 1950*. The Bill was passed in 2016 and became the *Biodiversity Conservation Act 2016*. Under the Act, all species of plants and animals are protected. The Act shares two 'objects', or objectives, with the international Convention on Biological Diversity, a 1993 multilateral treaty: to conserve and protect biodiversity components (such as ecosystem diversity, species diversity and genetic diversity) and to promote the ecologically sustainable use of biodiversity components. The Act also includes provision for the protection of threatened species and ecological communities; collapsed ecological communities; threatening processes and critical habitat; and the control of environmental pests.

Although biodiversity conservation is the primary purpose of the Act, it contains provisions dealing with bioprospecting. Clause 256 envisages the promulgation of regulations granting bioprospecting licenses subject to a condition "authorising bioprospecting activity that requires the licence holder to enter into an arrangement with the CEO or another person for the sharing of profits". The CEO is defined in the Act to mean the chief executive officer of the Department of Conservation and Land Management (now the Department of Biodiversity, Conservation and Attractions). "Bioprospecting activity" is defined as an "activity involving or related to the taking of fauna or flora for the purpose of identifying, extracting or recovering biological resources".[18]

A review of this Bill by the Environmental Defender's Office of Western Australia was critical of the fact that the Bill did not include as one of its objectives "to ensure the fair and equitable sharing of benefits resulting from the use of genetic resources/bioprospecting".[19]

There is huge potential to build new industries based on discoveries made using biological resources.[20] An ethical and sustainable bioprospecting industry has the potential to diversify the Western Australian economy, generate employment, and develop capability in high-tech areas as chemical screening and drug development.[21] However, existing intellectual property laws in Australia offer limited scope for the recognition of Indigenous peoples' rights regarding biodiversity-related knowledge and practices. Similarly, native title, heritage and environmental laws and policies provide insufficient means for addressing these rights.

The challenge is to protect Indigenous peoples' rights and knowledge while also conserving biological diversity. Having ratified the Convention on Biological Diversity in 1993, Australia must introduce initiatives that specifically recognise and protect the rights of Indigenous peoples to their biodiversity-related knowledge, innovations and practices.[22] Such initiatives would also provide opportunities to protect Indigenous knowledge practices and innovations related to biodiversity and to introduce measures for equitable sharing of benefits with traditional knowledge holders.

Wildfire on Mitchell Plateau.

CHAPTER 11

Managing threats to Kimberley rainforests

The Kimberley rainforests and their resident flora and fauna face threats from humans, animals and the environment. Managing these threats is thus an important factor in protecting and conserving the biodiversity of this region.

HUMAN DEVELOPMENT

To date, the north-west Kimberley has not been heavily affected by human development. However, the Kimberley has vast agricultural, aquacultural, horticultural, pastoral and geological resources. Mining tenements cover about 7.5% of the region, and include oil, gas, and rich mineral deposits of bauxite, copper, nickel, silver, lead, zinc, mineral sands, heavy rare earth elements (dysprosium, lutetium and terbium), potash and diamonds.[1] Gross value (sales) reported by the Kimberley Development Commission (a statutory authority of the Western Australian Government) generated by mining is $171-177 billion.[2,3]

The major agricultural activities of the Kimberley include horticulture (market gardening and fruit production) and the production of sugar cane, peanuts, sandalwood and cotton. There are 40 pastoral leases over Crown land across the Kimberley, which give the lessee the right to graze authorised livestock (such as sheep, cattle and goats) on natural vegetation. Marine aquaculture industries include pearling and the farming of marine finfish species that naturally occur within the Kimberley.[4]

The commercialisation of natural resources in the Kimberley has raised concerns about environmental impacts, including the impacts on natural systems, human health, scenery, recreational enjoyment, and other ecosystem services. Exploiting these resources could also have adverse effects on the biota of the region and present an ever-increasing challenge for conserving biodiversity in the Kimberley.[5,6]

TOURISM

Improved road conditions in the late 1970s and early 1980s (such as the grading and resurfacing of the Gibb River Road) encouraged more tourism in the Kimberley. By 2017, the region hosted some 593,000 visitors each year.[7] In 2021, the Broome-Cape Leveque Road – previously referred to by locals as 'Corrugation Road' due to its hazardous driving conditions – was sealed, thus increasing accessibility. With more people wanting to visit Cape Leveque, north of Broome, for recreation and camping, there is a real threat to the coastal monsoon rainforest patches on the Dampier Peninsula.

Visits to the Kimberley coast by sea are also increasing. When I travelled up the coast on a navy patrol boat in 1977, we observed only two yachts between Broome and Darwin. By 2006, at least 28 companies were operating tours on 30 vessels along the coast between Broome and Wyndham.[8] In 2013, some 300,000 tourists visited the coast, generating expenditure of more than $300 million.[9] Broome had been a port of call on extended itineraries for large cruise lines, with 37 ships scheduled to visit in 2020.[7] (Cruise ships were banned from visiting Australian ports from March 2020 to April 2022 due to the COVID-19 pandemic, but the Western Australian Government agreed that larger vessels could return to its ports from October 2022.) These cruises all involve excursions to remote sites along the Kimberley coast, including trips to fragile reef systems (such as Montgomery Reef) and along rivers and creeks lined with dense mangrove forests.[10] Tour boats and cruise ships produce bow waves that erode river banks, and the heavy scouring of sediment eventually causes mangrove trees to uproot and die. Increased tourism also produces waste and litter, which can pollute estuarine waters and damage the health of marine species.[11]

FIRE

The North Kimberley is recognised as a stronghold for declining native mammals and as an National Biodiversity Hotspot for endemic plants, snails, reptiles, frogs and mammals (one of only 15 such

hotspots recognised by the Australian Government).
To maintain this biodiversity, the Western
Australian Government developed the Kimberley
Science and Conservation Strategy in 2011.[12] An
important part of the strategy is the Landscape
Conservation Initiative being implemented through
partnerships between the former Department
of Parks and Wildlife, now the Department of
Biodiversity, Conservation and Attractions, and
Traditional Owners, pastoralists and other key
land managers including the Australian Wildlife
Conservancy. This initiative aims to:

> "Retain the current near pristine biodiversity
> and landscape values of the north-west
> Kimberley by preventing significant
> impacts from introduced animals, weeds,
> inappropriate fire regimes and other
> identified threats."[12]

European settlement in the Kimberley brought large
changes to the ecology and fire regimes as well as
a complete decline in traditional Aboriginal land
management practices, particularly burning.

There is little doubt that too-frequent, extensive
and hot fires have had a strong influence on the
localised distribution and boundary characteristics
of monsoon rainforests across northern Australia.
The surrounding tropical savannas constitute the
world's most fire-prone biome, and they support
a biota with a high degree of resilience to fire. The
fires occurring at the end of the dry season and the
onset of the wet season in this region are likely
natural, as the frequency of periodic lightning
strikes is high in October and November, little rain
has fallen, and most of the ground vegetation is
very dry and flammable. Periodic large fires are a feature of extensive, fire-prone
landscapes; not only are the landscapes highly resilient to them, but also the fires
can be important in promoting biodiversity.[13,14] The size and shape of persisting
patches often reflect the level of protection offered by surrounding landforms and
vegetation. Patches surrounded by rock outcrops, creeks and rivers or bordered
by ocean are burned less frequently than those adjacent to savanna woodlands,
with their abundant seasonally dry grasses. Regular or intense wildfires break
the dense cover of foliage that helps to maintain the moisture balance and local
environment for rainforest species (Figs 155-159).

FIGURE 155

An ash trail from a burnt
tree, leading into a monsoon
rainforest patch at the Lone
Dingo rainforest patch on
Mitchell Plateau.

FIGURE 156

Large, uncontrolled wildfires
burn across the Kimberley
landscape, usually late in the
dry season. Every year,
around 40% of the Kimberley
region burns.

FIGURE 157 (LEFT, TOP)

The remnants of a savanna woodland after fire, at the edge of the Crusher Thicket monsoon rainforest patch on Mitchell Plateau.

FIGURE 158 (ABOVE)

Repeated hot fires can cause enormous damage to savanna and monsoon rainforest vegetation.

FIGURE 159 (LEFT, MIDDLE)

Joe Raudino on a 1993 Landscope Expedition to Mitchell Plateau, where fire had intruded into a monsoon rainforest patch. The fire can smoulder for some time and then reignite, causing more damage.

FIGURE 160 (LEFT, BOTTOM)

Cattle trampling and grazing damaged this monsoon rainforest patch on Mitchell Plateau.

Although officially banned by government regulations in many countries, the use of fire plays an important role in a range of savanna management and biodiversity conservation applications in Australia.[15] Prescribed burns are conducted in the late wet and early dry seasons to create a mosaic of burnt and unburnt areas across almost 200,000 square kilometres of the North Kimberley. The resulting effect is a reduction in the amount of available fuel and fewer large, intense and damaging fires later in the dry season. Around one-third of the Kimberley is burnt every year, a rate that is detrimental to all fire-sensitive species.[16]

By managing landscape-scale threats, we can also protect many of the plants and vegetation communities, including monsoon rainforest.[17] An examination of historical and recent aerial photos from 1949, 1969 and 2005 found that the rainforest patches on Mitchell Plateau expanded by 9% over this period, despite severe wildfire events.[18] Increasing rainfall in northern Australia since the 1940s has been a key driver of rainforest expansion.[19] In that same period, monsoon rainforests on the Bougainville Peninsula expanded by 69%, due primarily to the absence of cattle on the Peninsula and less frequent fire.[18] In the 1980s, a boundary fence was constructed across a narrow isthmus on the Bougainville Peninsula (now part of the Uunguu Indigenous Protected Area) to keep feral cattle and donkeys from the area. This is helping to conserve one of the largest continuous patches of monsoon rainforest in the Kimberley. The cattle-free and rarely burnt Bougainville Peninsula may be important for biodiversity, as it offers more long-unburnt habitats for large numbers of small to medium-sized threatened mammals and fire-sensitive plant species rarely found elsewhere in northern Australia.[18]

The expansion of some rainforests into savannas suggests that rainforest saplings could have traits that enable them to survive in the savanna environment, including the ability to recover from infrequent fires. Researchers have hypothesised that rainforest expansion has occurred in the North Kimberley in response to the wetting trend in northern Australia over the past century, together with increasing atmospheric carbon dioxide, but that the expansion is strongly influenced by the combined effects of fire and feral cattle, which may also be pivotal in maintaining sharp floristic and structural distinctions between rainforests and savannas.[20,21]

Cattle use rainforest patches for shade, but they trample the understorey and open patches up to invasion by both flammable savanna grasses and introduced grasses (Fig. 160).[12] Climate change and the resulting severe heatwaves could put further pressure on rainforest patches by promoting more severe fires. As temperatures increase, more cattle will be driven to seek shade.[22] Moist climate, infrequent fire and geology are important stabilising factors that allow rainforest fragments to persist as islands in a sea of savanna.[21]

Concerted efforts to better manage fire and feral animals (cattle in particular) by the Department of Biodiversity, Conservation and Attractions, the Kimberley Foundation (on Theda and Doongan pastoral stations) and the Australian Wildlife Conservancy are improving the status of threatened and other species in the Kimberley. Several fire projects in the Kimberley have made great inroads in terms of arresting the boom-and-bust fire regimes that dominated the North Kimberley before 2008.[23] Prescribed burning has reduced late-season fires by nearly 50%, resulting in a 55% increase in vegetation patchiness (for ≥3-year-old vegetation). These projects include ECOFire run by the Australian Wildlife Conservancy on Central Kimberley pastoral lands and the Fire Abatement projects run by four Native Title groups (Wunambal Gaambera, Dambimangari, Willinginn and Balanggarra) together with the Kimberley Land Council. Indigenous ecological knowledge and science are increasingly combined to inform land management decisions. Many of these projects generate Australian carbon credit units through application of Australia's formal savanna burning methodology.[15,24,25]

Monitoring and evaluation have found a very strong relationship between fire frequency and mammal abundance and diversity.[26] Areas that are burnt more than once every two years had the fewest animals, whereas sites that are burnt less than once every five years (such as parts of Prince Regent National Park) had the most. There are still areas of high cattle density and high fire frequency, such as the King Edward River area, where improvements to ecosystem management and biodiversity conservation are needed.[27] Cattle are an important source of food for Aboriginal people in the Kimberley, as in other Aboriginal communities in Northern Australia. Traditional Owners are developing programs that balance their food needs with conservation objectives. Rainforests are generally considered as refugia from the extensive savanna fires of the Kimberley and are thus critical to providing an avenue for recolonisation after fire.

FERAL ANIMALS

Feral cattle and pigs cause damage to rainforests through trampling and destruction of vegetation. Control of large feral herbivores has yet to measurably reduce their impact on the landscape, and ongoing monitoring is needed to evaluate changes in their impact (Fig. 161).[23]

Since it was first found in Western Australia in 2009, the invasive Cane Toad (*Rhinella marina*) has become established in the East and North Kimberley. The eggs, tadpoles and adults are toxic to a wide range of Kimberley wildlife. Cane Toads also threaten native species through predation and competition for habitat. The species is listed by the International Union for Conservation of Nature as one of the world's worst invasive alien species. There is minimal natural predation of Cane Toads in Australia, and their movement across the Kimberley cannot be stopped using any of the methods currently available.[28]

FIGURE 161 (ABOVE)

Feral cattle have destroyed monsoon rainforest patches in the Kimberley.

FIGURE 162 (LEFT)

Numbers of *Boorroo, Aamba* or Agile Wallaby (*Macropus agilis*) have declined in the monsoon forest next to the Broome townsite due to harassment from domestic dogs.

Rapid urban expansion at Broome has resulted in domestic dogs harassing wildlife.[ii] This in turn has led to a decline in the number of Agile Wallabies feeding in the monsoon rainforest at Minyirr Coastal Park, located behind the sand dunes at Cable Beach (Fig. 162).

WEEDS

Weeds are a major threat to Australia's biodiversity, as they contribute to hotter, more frequent fires, modify soil characteristics and compete directly with native plant species.[29] In many countries, non-native species are second only to climate change as the biggest threat to native biological diversity.[30] Their impacts on agriculture amount to millions of dollars annually.[31]

ii. This observation was made by Tim Willing, a member of the Broome Botanical Society and a Broome resident for many years.

FIGURE 163 Siratro or Purple Bean (*Macroptilium atropurpureum*) is the most serious weed affecting monsoon rainforest on the Dampier Peninsula. The fast-growing creeper smothers vegetation and provides biomass for hotter fires.

FIGURE 164 Stinking Passionflower (*Passiflora foetida*), an aggressive climber, is one of the biggest threats to the highly fragmented rainforest patches of the Kimberley. The epithet *foetida* ('stinking' in Latin) refers to the strong smell emitted by damaged foliage.

In the coastal monsoon vine thickets of the Dampier Peninsula, the most serious weed is Siratro or Purple Bean (*Macroptilium atropurpureum*). This fast-growing, deep-rooted legume has a climbing stem that grows over trees and shrubs, smothering most species and creating a fire hazard (Fig. 163). A native species of North, Central and South America, it was introduced into northern Australia as a fodder plant but has become extremely invasive, especially along river systems. Siratro is considered an environmental weed in Queensland, the Northern Territory, northern Western Australia and northern New South Wales.[32]

More widespread in the Kimberley is the Stinking Passionflower (*Passiflora foetida*), which smothers rainforest patches and can draw fire into the canopy, causing tree deaths (Fig. 164). This weed is thought to be one of the biggest threats to the highly fragmented rainforest patches in the Kimberley. It is currently listed under the *Environment Protection and Biodiversity Conservation Act* as being among the main weed threats to the vine thickets of the Dampier Peninsula.[33] The Stinking Passionflower forms 'sails' in the upper canopy, which provide fuel for fires and can cause trees to be uprooted during high winds, such as those associated with cyclones and other tropical storms. Its seeds are spread by birds and other animals. Genomic research has shown that the majority of Stinking Passionflower samples collected from Western Australia and northern Australia are likely to have originated in Ecuador or Peru, but how and when these introductions occurred is not known.[34]

On Mitchell Plateau, the introduced Grader Grass (*Themeda quadrivalvis*) and Annual Mission Grass (*Cenchrus pedicellatus* subsp. *unispiculus*) are of concern. Grader Grass is a declared pest in Western Australia, while Annual Mission Grass is considered an invasive pasture grass by the Australian Government. Although Annual Mission Grass has been eradicated on Mitchell Plateau, it is still present at Kalumburu and could be reintroduced. Because Annual Mission Grass can grow in shady areas, it helps to fuel fires beneath shrubs and trees that wouldn't otherwise burn so easily. As a result, the weed could potentially have a high impact in the Kimberley and is a key threat to biodiversity in Northern Australia.[35] Both Grader Grass and Annual Mission Grass create fuel loads that are substantially greater than those of most native grasses.[36]

Also on Mitchell Plateau, the introduced Paddy's Lucerne or Arrowleaf Sida (*Sida rhombifolia*) is regarded as a significant environmental weed. It is difficult to eradicate and is well-established around infrastructure. It is also considered an environmental weed in the Northern Territory, Queensland and New South Wales.[iii]

---

iii. This information was provided by Tom Vigilante in a personal communication.

DISEASE

There is the risk that feral herbivores could spread foot-and-mouth disease in northern Australian cattle herds and could carry *Salmonella*, the most frequently reported cause of food-related illness in humans.[37] An increase in international tourism, to islands such as Bali, has been linked to the spread of non-native *Salmonella* serovars to humans and animals, which suggests that densely populated islands in Asia could act as staging points for the transmission of pathogenic microbes between tropical and temperate regions.[38] Outbreaks of foot-and-mouth disease across Indonesia, including in Bali, in mid-2022 prompted the Australian Government to provide vaccines, funding and technical support to help Indonesia control the disease and prevent its spread.[39]

Another threat to rainforest plants and the surrounding savanna vegetation in the Kimberley is Myrtle Rust (*Austropuccinia psidii*), a pathogenic fungus that was detected in the east Kimberley in August 2022 (Fig. 165). Native to South America, it was first detected in New South Wales in 2010 and has now spread to all states and territories except South Australia. Myrtle Rust infects plants in the myrtle family (Myrtaceae), such as *Backhousia*, *Corymbia*, *Eucalyptus*, *Melaleuca*, *Verticordia*, *Xanthostemon* and *Syzygium* species, destroying new growth and soft tissue and eventually killing the plant. In New South Wales and Queensland, around 40% of Myrtaceae genera are infected.[40] Myrtle Rust spores are easily spread via contaminated clothing, hair, skin, personal items, plant material and equipment, as well as by insect and animal movement and wind dispersal. The presence of Myrtle Rust in the Kimberley will have dire consequences for the region's biodiversity, and monitoring its spread needs to be an urgent priority for state and federal governments.

The Great Bowerbird (*Chlamydera nuchalis*) is found on the edges of monsoon rainforest. *Image: Rini Kools/Shutterstock*

CHAPTER 12
Conserving the biodiversity of Kimberley rainforests

Mounting evidence of the Kimberley as a historical and ancient centre of refugia warrants action from scientists, government, conservation agencies, Indigenous landholders and local communities to protect and conserve its unique biota, as well as the processes responsible for generating and sustaining it. In October 2014, the Western Australian Government Minister for Environment and Heritage, the Hon. Albert Jacob, stated that "Biodiversity surveys assist Kimberley land management".[1] He added: "Surveys of animals and plants in key locations throughout the Kimberley are providing important information to help conserve the region's biodiversity."

The north-west Kimberley is widely regarded by researchers as one of the world's last great botanical frontiers. However, such a diverse and extensive region as the Kimberley also poses a formidable challenge to scientists trying to determine what occurs where – the first of many questions that need to be addressed in effective research and conservation. Most of the scientific research undertaken in the region has been directed toward answering this pivotal question.

The high rainfall, near-coastal region of the north-west Kimberley has experienced no known mammal extinctions to date,[2] but several species with formerly wider distributions have now contracted to this area. This includes three of the Kimberley's four endemic mammalian species: the *Yilangal* or Scaly-tailed Possum (*Wyulda squamicaudata*), the *Monjon* (*Petrogale burbidgei*) and the Kimberley Rock Rat (*Zyzomys woodwardii*).[3]

The Kimberley is recognised as a centre of mega-diversity and micro-endemism, with the distributions of taxa undoubtedly shaped by changing climates, landscape and geomorphology. However, the lack of fine-scale sampling across the Kimberley and surrounding bioregions has hindered a greater understanding of the

The Magnificent Tree Frog (*Litoria splendida*), which is found only in the north-west Kimberley and the Northern Territory, lives in caves and rock crevices in deep gorges. *Image: Mark Cowan*

biogeographical history of the region. Flooding in the wet season results in roads being closed for many months at a time. Much of the region is accessible only by helicopter or by boat from the sea. Limited access to this topographically rugged and remote area has largely restricted comprehensive assessments of its biodiversity, and only in recent years has the compositional complexity of its flora and vertebrate fauna been fully appreciated.[4]

The major impediments to carrying out fine-scale sampling in the Kimberley are its remoteness, limited access and the lack of funding to carry out long-term monitoring. Cost-effective solutions for monitoring rainforest patches include technologies such as satellite monitoring, remote instruments (including camera traps), remotely piloted aircraft (RPAs, or drones) and enlisting the power of citizen science to help fund research – as demonstrated by the very successful Landscope Expedition to Mitchell Plateau area in 2002.[5] Most important are the collaborative projects currently being undertaken with Traditional Owners.

Scientists have looked to islands for insights into the forces that shape biological diversity, which include dispersal, invasion, competition, adaption and extinction. A predictive understanding of extinction might be obtained from island biogeography, since refugia behave as islands for species that are dependent on natural habitat.[6] The importance of dispersal is increasingly being recognised for its fundamental role in the generation of biodiversity on islands.[7] Much of island biogeography, and much of contemporary ecology, deals

The King Brown Snake (*Pseudechis australis*) is widespread in the Kimberley and can be found in monsoon rainforest. *Image: Norm McKenzie*

with changing distributions, patterns and interactions of flora and fauna communities.[8] Until recently, most references to diversity concerned species-level diversity and not the integrity of biological communities. Resource management now focuses increasingly on the community level. These analyses and principles are directed not merely at saving endangered species, but rather at keeping the full complement of species and assemblages from becoming endangered.[9]

Fewer than 10% of the 1500 rainforest patches in the Kimberley have been studied intensively. From these, we have a reasonable understanding of the flora and vertebrate fauna assemblages, but all patches have their own unique mix of species. What we do not have is a good understanding of the non-vascular plants (lichens, bryophytes and fungi) and the majority of invertebrate groups, particularly the less mobile species. However, we know from research into land snails and earthworms that species can be endemic within their own patch of rainforest. It is essential that further studies of monsoon rainforest are undertaken to document long-term temporal changes, the similarities and differences between isolated patches, and the ephemeral species that occur only during the wet season.

The Landscape Conservation Initiative established by the Western Australian Government in 2011 as part of the Kimberley Science and Conservation Strategy aims to retain and enhance the high biodiversity and landscape values of the North Kimberley, an area of more than 65,000 square kilometres. Some of the key achievements of this strategy have been the removal of feral cattle, pigs, cats and horses from areas with high biodiversity values and the use of prescribed burning in the early dry season to greatly reduce destructive late-season fires.[10] Some Kimberley monsoon rainforests are located within national parks or other conservation reserves and are therefore afforded some protection. Others are located within Aboriginal reserves, and it may be appropriate to establish these sites as Indigenous Protected Areas (IPAs). An IPA is an area of land or sea that has been voluntary declared to be a protected area by its Traditional Owners. IPAs are managed by Indigenous peoples according to international guidelines and are recognised by Commonwealth, state and territory governments as part of Australia's National Reserve System. The establishment of IPAs could avoid past practices where Indigenous peoples' rights and interests were either overlooked or forced to fit within the structures and processes promoted by government.[11]

The focus of conservation should be to protect monsoon rainforest patches across the Kimberley. They require management across different tenures, both within and outside any reserves, that might yet be established. Remote does not always imply pristine. At first glance, the size of the ancient, often massive landscapes in the Kimberley can mask their fragility.[12] In the past, much of the management of this vast area was one of 'benign neglect'. The monsoon rainforests of the Dampier Peninsula are protected under state and federal biodiversity legislation. In 2011, the West Kimberley was included on the National Heritage List, pursuant to section 324JJ of the *Environment Protection and Biodiversity Conservation Act 1999*.[13] However, this legislation provided limited protection to the larger, more complex patches in the north-west Kimberley. Monsoon rainforests of the northern Kimberley coast and islands were specifically cited for their outstanding heritage value to the nation for their evolutionary refugial role that has resulted in high invertebrate richness and endemism. The National Heritage listing does not change land ownership and does not affect Native Title. Management of places on the National Heritage List remains with the current landowner or manager. The inclusion of the West Kimberley on the National Heritage List helps ensure that heritage values are part of decision-making for this area, and that heritage protection is balanced with the social and economic aspirations of the Kimberley community.[14]

In Western Australia, only the monsoon thickets of the Dampier Peninsula and the Theda Swamp, Walcott Inlet and Roe River rainforests and swamp forests are listed by the federal government as threatened ecological communities. Rainforests at Point Spring,

Lichen (*Ramalina subfraxinea*) in monsoon rainforest. *Image: Tim Willing*

Napier and Ningbing Ranges as well as Vegetation Association 717, an ecological community defined as mixed tropical deciduous forest, are listed as priority ecological communities.[15] In Western Australia, the Environmental Protection Authority (EPA) has issued a guidance statement for the protection of mangroves along the Pilbara coastline.[16] Guidance statements provide advice on the minimum requirements for environmental management that the EPA expects to be met when it considers proposals that are likely to have potentially significant effects on the environment. As yet, no such guidance statement has been prepared for Kimberley monsoon rainforest patches.

Roger Hnatiuk and Karl Pirkopf in dense Cane Grass (*Sorghum amplum*) fringing monsoon forest on Mitchell Plateau in 1986.

CHAPTER 13
Establishing a tropical biodiversity research centre in the Kimberley

After nearly five decades of investigating the complexity of Kimberley monsoon rainforests, savanna woodlands and wetlands, it has become apparent to me that there is a fundamental need to continue to investigate their biodiversity, biogeographic relationships and potential for bioprospecting, as well as ensuring their effective conservation and management. There is an urgent requirement to instil an appreciation of their natural and cultural values within the landscape and across the wider tropics. This can only be undertaken in partnership with Traditional Owners. For over 50 millennia, Indigenous peoples have developed an intimate and unique connection with Country. They have established distinct systems of knowledge, innovation and practice relating to the uses and management of biological diversity. As many publicly available records of Indigenous biocultural knowledge relate to the Kimberley, working with Traditional Owners presents an opportunity to develop a list of key regions based on their biological and cultural importance.[1]

A dedicated research centre is necessary for conducting integrated tropical studies in the Kimberley. Securing the agreement and cooperation of Traditional Owners is critical. Large areas of the Kimberley are private lands managed by Traditional Owners and require special collecting permits and protocols administered by the Kimberley Land Council on behalf of the Traditional Owners or direct negotiation with the Traditional Owners themselves.[2] Conducting research with Traditional Owners is both respectful and appropriate and can lead to productive outcomes for both the researchers and Indigenous communities. Aboriginal Elders' knowledge of the natural environment comes from acute observation in the course of their daily lives.[3]

The Amax base camp on Mitchell Plateau in 1982, now the site of the Kandiwal community.

Approaches have been made in the past to the Western Australian Government for the establishment of such a centre. In October 1983, when Conzinc Riotinto of Australia planned to mothball its bauxite project on Mitchell Plateau, the Western Australian Naturalists' Club wrote to the responsible minister and requested that the government purchase the site and establish a tropical studies centre. The Club pointed out that Mitchell Plateau was an important location offering a diversity of habitats, an all-weather airstrip and access to the sea at Port Warrender. The government considered the request but, due to what it described as serious financial hurdles, took no action, and the opportunity was lost.[4] In December 1983, I raised the need for a research facility in the Kimberley to help build up a continental perspective of Australia's rainforests.[5] Again, no action from the government was forthcoming.

Almost 40 years later, the need for a permanent research centre in the Kimberley has only become more urgent. In that time, more traditional knowledge, wisdom and folklore has been lost.[6] We have missed the opportunity to study the effects of climate change on the ecosystem. We have failed to describe and study new species, many of which may be restricted to the area and may be rare or threatened.

Establishing a research centre in the region would help to deepen our understanding not only of the systematics and ecology of Kimberley flora and fauna, but also of the region's biogeographic connection to South-East Asia, New Guinea and the Cape York Peninsula.

River weed *Malaccotristicha australis* grows on submerged rocks in the rapids at Mitchell Falls. When water levels drop in the dry season, its flowers open above water. It is also found in the Northern Territory, while *M. malayana* is found in Peninsular Malaysia and Peninsular Thailand.

Greater knowledge of what species exist, where they occur, and their ecological requirements for survival would support the development and implementation of improved fire strategies. It would also enable the systematic mapping and cataloguing of the region's biological resources, allowing us to place economic and environmental value on our biota and Indigenous knowledge.

From a practical perspective, a permanent research centre would enable year-round access to remote areas of the Kimberley that are otherwise unreachable, particularly during the wet season (the best time of year for collecting fertile plant material), as well as providing adequate facilities for drying and preserving specimens in humid conditions. An on-ground presence would also make monitoring of plant and animal diseases such as Myrtle Rust and foot-and-mouth disease much timelier and more effective.

A permanent centre in the Kimberley would also facilitate longer-term research. The status and trends of monsoon rainforest biodiversity could be tracked more closely over longer periods, leading to an increased understanding of ecosystem function and cumulative pressures.[7] Researchers would also have more opportunities to forge stronger, long-lasting relationships with Traditional Owners and learn from their innate understanding of Country. As a biodiversity hotspot, the Kimberley provides an ideal location for undertaking collaborative, globally relevant and applied research to understand, conserve and responsibly manage monsoon rainforests and savanna woodlands.

ACKNOWLEDGEMENTS

This book is based on a paper published in the *Journal of the Botanical Research Institute of Texas* in 2018, and has been updated and expanded to include recent advances in rainforest research and taxonomic changes to plant names. I would particularly like to acknowledge and thank the following people for their assistance, guidance and advice in the preparation of the original paper and this book.

The Traditional Owners of the Kimberley, who have been extremely generous over many decades in conducting me through Country and sharing their vast wisdom, knowledge, culture and beliefs, which has allowed me to have a far greater appreciation of this ancient landscape. Permission was sought from and granted by Traditional Owners to record the language, images and information contained in this book, including references to Indigenous traditional knowledge and to biological resources (plants and animals). They are published for the purposes of education and knowledge sharing.

Professor Ric How (former Curator of Biogeography, Western Australian Museum) for his invaluable discussions, comments and advice on the fauna of the Kimberley; Vince Kessner for his devotion and zeal to Kimberley snail collecting and for providing outstanding snail images; Ron Johnstone OAM (Curator of Ornithology, Western Australian Museum) for advice on the avifauna of rainforests and for access to images; Norm McKenzie AM (former Senior Principal Research Scientist, Department of Parks and Wildlife) for sharing his encyclopaedic knowledge of Kimberley rainforests and providing several images; John Dell PSM for reviewing the manuscript and making valuable comments; Dr Matt Barrett and Dr Russell Barrett (Kings Park and Botanic Garden) for advice on rainforest fungi and for providing images, respectively; Dr Tom Vigilante (Bush Heritage Australia) and Dr Alex George AM (Murdoch University) for their valuable comments on the original publication; Dr Frank Kőhler (Australian Museum) for advice and research papers on the biogeography of rainforest snails; and Brian Carter, Kevin Coate, David Haig, Jiri Lochman (Lochman Transparencies), Joe Smith and Tim Willing for kindly providing images. Louise Beames (Environs Kimberley) kindly provided the distribution map of the monsoon forest on the Dampier Peninsula.

I would also like to thank Dr John Huisman (Curator) and the staff of the Western Australian Herbarium for access to the plant collections housed there; my colleagues (past and present) at the Western Australian Museum for their advice, discussions and good company over many field trips to the Kimberley; and Dr Karl-Heinz Wyrwoll (Senior Research Fellow, University of Western Australia) for advice on the climatic history of the Kimberley. I am indebted to Dr Bernie Hyland (former rainforest botanist with the Australian National Herbarium, CSIRO) for his willingness to share with me his vast knowledge of rainforest plants on Kimberley field trips, and I pay homage to his mastery of the brush hook in creating ready access to rainforest patches. I also acknowledge his foresight in developing Australian Tropical Rainforest Plants, an interactive identification and information system that covers all the trees, shrubs and vines across northern Australia's rainforests.

I am also extremely indebted to my friends at Biota Environmental Sciences, particularly Paul Sawers (GIS/Spatial Data Manager) for his cartographic skills in producing several of the figures and maps that appear in this book and Michi Maier (Principal Botanist/Director) for reviewing the manuscript and making valuable comments. Artist Rob Fleming kindly prepared the colour version of the rainforest structure diagram. Dr Mike Donaldson (Wildrock Publications) graciously granted me permission to reproduce his geology map of the Kimberley.

Suggestions from Professor Jeremy Russell-Smith (Charles Darwin University) and Dr Stefania Ondei (University of Tasmania) also helped me improve the original publication considerably. I am extremely grateful for the support and encouragement I received from Dr Barney L. Lipscomb, Editor-in-Chief, Botanical Research Institute of Texas, on the original publication.

I would like to sincerely thank Biota Environmental Sciences, Western Australian Naturalists' Club, and Western Australian Gould League at the Herdsman Lake Discovery Centre for financial support of the original publication.

Finally, I thank my wife, Dr Irene Ioannakis, for her patience during my absence on field work and for her assistance in putting this publication together.

Professor Kevin Kenneally AM is an Adjunct Professor in the UWA School of Agriculture and Environment and at the Nulungu Research Institute, Notre Dame University Australia.

Kevin became interested in rainforest more than 50 years ago when he was conscripted into the Royal Australian Army during the Vietnam War and posted to Queensland. At the time of his military call-up in 1967, Kevin was working in the Botany Department at UWA. The head of department, Professor Brian Grieve, gave him a letter of introduction to the directors of all the state botanical gardens and herbaria (where pressed and dried botanical specimens are housed for taxonomic study) in Australia. When on leave from the army, Kevin visited the Royal Botanic Gardens in Melbourne. On being posted to Brisbane, he visited the Queensland Herbarium, then at the City Botanic Gardens, where he first encountered rainforest plants. Army training exercises at the Jungle Warfare Training Centre at Canungra, located amid thick rainforest and steep, razor-back country 75 km south of Brisbane, gave him his first real field experience in dealing with jungle-like vegetation.

After discharge in 1969, Kevin returned to the Botany Department at UWA. In 1973, he joined the staff of the Western Australian Herbarium as a research botanist. In the same year, the State Government commenced a series of biological survey expeditions to the Kimberley, starting with an investigation of some of the Kimberley islands. Kevin joined one of these expeditions in 1974 to the Prince Regent River Nature Reserve (now Prince Regent National Park), where he encountered his first patch of Kimberley rainforest. The rest, as they say, is history.

An opportunity to conduct intensive studies of the rainforest patches at Mitchell Plateau arose in 1976 following the establishment of a mining camp there by Amax. A permanent campsite with an all-weather air strip and permission to use the company's facilities allowed Kevin to perform the first wet-season collecting in Kimberley rainforests. Between June 1987 and March 1989, Kevin was part of a team of scientists who conducted intensive surveys of Kimberley rainforests under the National Rainforest Program, funded by the Australian Government and coordinated by Norm McKenzie from what was then the Department of Conservation and Land Management. Around that time, Kevin had the good fortune to meet and work with Dr Bernie Hyland, a rainforest specialist with the Australian National Herbarium at CSIRO. Bernie's vast experience in north Queensland rainforests and his willingness to share his knowledge of rainforest plants was invaluable. Kevin's continued participation in surveys over the years has helped him to understand the complexity and value of the Kimberley rainforest.

(Opposite) Kevin Kenneally, National Service, 1968; (top) Kevin Kenneally at the Western Australian Herbarium in 1979; (middle) the Kimberley trinity Kevin Kenneally, Daphne Edinger and Kevin Coate on top of Oomarri (King George Falls); (bottom) Kevin Kenneally in the Kimberley. *Image: Daphne Edinger*

REFERENCES

INTRODUCTION

1. Nunez C (2019) Rainforests, explained. *National Geographic* website, 16 May

2. Grainger A (1980) The state of the world's tropical forests. *Ecologist* 10: 6-54

3. Koenen EJM, Clarkson JJ, Pennington TD, Chatrou LW (2015) Recently evolved diversity and convergent radiations of rainforest mahoganies (Meliaceae) shed new light on the origins of rainforest hyperdiversity. *New Phytologist* 207(2): 327-339

4. Kreft H, Jetz W (2007) Global patterns and determinants of vascular plant diversity. *Proceedings of the National Academy of Sciences USA* 104(14): 5925-5930

5. Russell-Smith J, Bowman DMJS (1992) Conservation of monsoon rainforest isolates in the Northern Territory, Australia. *Biological Conservation* 59(1): 51-63

6. Kenneally KF (1989) Checklist of the vascular plants of the Kimberley, Western Australia. Handbook no. 14. WA Naturalists' Club, Perth, Australia

7. Barrett RL (2015) Fifty new species of vascular plants from Western Australia - celebrating fifty years of the Western Australian Botanic Garden at Kings Park. *Nuytsia* 26: 3-20

8. Australian Government (2015) Our north, our future: white paper on developing Northern Australia. Commonwealth of Australia, Canberra, Australia

CHAPTER 1: TROPICAL RAINFORESTS

1. Baccini A, Walker W, Carvalho L, Farina M, Sulla-Menashe D, Houghton RA (2017) Tropical forests are a net carbon source based on aboveground measurements of gain and loss. *Science* 358(6360): 230-234

2. Exbrayat JF, Liu YY, Williams M (2017) Impact of deforestation and climate on the Amazon Basin's above-ground biomass during 1993-2012. *Scientific Reports* 7: 15615

3. Milman O (2018) Vehicles are now America's biggest CO_2 source but EPA is tearing up regulations. *The Guardian*, 1 January

4. Watts J (2017) Alarm as study reveals world's tropical forests are huge carbon emission source. *The Guardian*, 29 September

5. McKenzie NL, Johnston RB, Kendrick PG (eds) (1991) *Kimberley rainforests of Australia*. Surrey Beatty & Sons, Chipping Norton, Australia

6. Spracklen DV, Garcia-Carreras L (2015) The impact of Amazonian deforestation on Amazon basin rainfall. *Geophysical Research Letters* 42(1): 9546-9552

7. Schimper AFW (1903) *Plant-geography upon a physiological basis*. Clarendon Press, Oxford, UK

8. Eamus D (1999) Ecophysiological traits of deciduous and evergreen woody species in the seasonally dry tropics. *Trends in Ecology & Evolution* 14(1): 11-16

9. Beard JS (1976) The monsoon forests of the Admirality Gulf, Western Australia. *Vegetatio* 31: 177-192

10. Clayton-Green KA, Beard JS (1985) The fire factor in vine thicket and woodland vegetation of the Admiralty Gulf region, north-west Kimberley, Western Australia. In: Ridpath MG, Corbett LK (eds) *Ecology of the wet-dry tropics*, pp 225-230. Ecological Society of Australia, Melbourne, Australia

11. Kenneally KF, Keighery GJ, Hyland BPM (1991) Floristics and phytogeography of Kimberley rainforests, Western Australia. In: McKenzie NL, Johnston RB, Kendrick PG (eds) *Kimberley rainforests of Australia*, pp 93-131. Surrey Beatty & Sons, Chipping Norton, Australia

12. Webb LJ (1959) A physiognomic classification of Australian rainforests. *Journal of Ecology* 47: 551-570

13. Russell-Smith J, Lee AH (1992) Plant populations and monsoon rain forest in the Northern Territory. *Biotropica* 24(4): 471-487

14. Bowman DMJS, Wilson BA, McDonough L (1991) Monsoon forests in northwestern Australia. I. Vegetation classification and the environmental control of tree species. *Journal of Biogeography* 18: 679-686

15. Webb L, Tracey JG (1981) Australian rainforests: patterns and change. In: Keast A (ed) *Ecological biogeography of Australia*, pp 605-694. Dr W Junk, The Hague, The Netherlands

16. Bradley AJ, Kemper CM, Kitchener DJ, Humphreys WF, How RA (1987) Small mammals of the Mitchell Plateau Region, Kimberley, Western Australia. *Australian Wildlife Research* 14(4): 397-413

17. Brockman JG (1880) Journal of an exploring trip from Beagle Bay to the Fitzroy River and back again. *The West Australian*, 28 May

18. Hill E (1933) The legion of the lost ones. *Chronicle* (Adelaide), 29 June

19. Gardner CA (1944) The vegetation of Western Australia. Presidential Address, 1942. *Journal of the Royal Society of Western Australia* 28: xi-lxxxviii

20. Baur GN (1968) The ecological basis of rainforest management. Government Printer, Sydney, Australia

21. Bowman DMJS (2000) *Australian rainforests: Islands of green in a land of fire.* Cambridge University Press, Cambridge, UK

22. Zimmermann F (1987) *The jungle and the aroma of meats: an ecological theme in Hindu medicine.* University of California Press, Berkeley, CA, USA

23. Slater C (2003) *In search of the rain forest.* Duke University Press, Durham, NC, USA

24. Richards PW (1952) *The tropical rainforest: An ecological study.* Cambridge University Press, London, UK

CHAPTER 2: THE KIMBERLEY

1. Finch D, Gleadow A, Hergt J, Heaney P, Green H, Myers C, Veth P, Harper S, Ouzman S, Levchenko VA (2021) Ages for Australia's oldest rock paintings. *Nature Human Behaviour* 5: 310-318

2. Di Nezio PN, Timmermann A, Tierney JE, Jin F-F, Otto-Bliesner B, Rosenbloom N, Mapes B, Neale R, Ivanovic RF, Montenegro A (2016) The climate response of the Indo-Pacific warm pool to glacial sea level. *Paleoceanography and Paleoclimatology* 31(6): 866-894

3. Williams AN, Turney C, Cadd H, Shulmeister J, Bird M, Thomas Z (2020) The last ice age tells us why we need to care about a 2°C change in temperature. *The Conversation*, 25 February

4. Ali JR, Heaney LR (2021) Wallace's line, Wallacea, and associated divides and areas: history of a tortuous tangle of ideas and labels. *Biological Reviews of the Cambridge Philosophical Society* 96(3): 922-942

5. Vigilante T, Ondei S, Goonack C, Williams D, Young P, Bowman DMJS (2017) Collaborative research on the ecology and management of the 'Wulo' monsoon rainforest in Wunambal Gaambera Country, North Kimberley, Australia. *Land* 6(4): 68

6. Murphy BP, Bowman DMJS (2012) What controls the distribution of tropical forest and savanna? *Ecology Letters* 15(7): 748-758

7. Staver AC, Archibald S, Levin SA (2011) The global extent and determinants of savanna and forest as alternative biome states. *Science* 334(6053): 230-232

8. Bradshaw CJA, Norman K, Ulm S, Williams AN, Clarkson C, Chadoeuf J, Lin SC, Jacobs Z, Roberts RG, Bird MI, Weyrich LS, Haberle SG, O'Connor S, Llamas B, Cohen TJ, Friedrich T, Veth P, Leavesley M, Saltré F (2021) Stochastic models support rapid peopling of Late Pleistocene Sahul. *Nature Communications* 12: 2240

9. Balme J (2013) Of boats and string: the maritime colonization of Australia. *Quaternary International* 285: 68-75

10. O'Connor S (1995) Carpenters Gap Rockshelter 1: 40,000 years of Aboriginal occupation in the Napier Ranges, Kimberley, WA. *Australian Archaeology* 40(1): 58-59

11. Balme J (2000) Excavations revealing 40,000 years of occupation at Mimbi Caves, south central Kimberley, Western Australia. *Australian Archaeology* 51(1): 1-5

12. O'Connor S, Veth P (2006) Revisiting the past: changing interpretations of Pleistocene settlement subsistence and demography in northern Australia. In: Lilley I (ed) *Archaeology of Oceania: Australia and the Pacific Islands, pp 31-47. Blackwell*, Oxford, UK

13. O'Connell JF, Allen J, Williams MAJ, Williams AN, Turney CSM, Spooner NA, Kamminga J, Brown G, Cooper A (2008) When did Homo sapiens first reach Southeast Asia and Sahul? *Proceedings of the National Academy of Sciences USA* 115(34): 8482-8490

14. Russell D (2004) Aboriginal-Makassan interactions in the eighteenth and nineteenth centuries in northern Australia and contemporary sea rights claims. *Australian Aboriginal Studies* 1: 3-17

CHAPTER 3: INDIGENOUS BIOCULTURAL KNOWLEDGE OF KIMBERLEY RAINFORESTS

1. Veth P, Myers C, Heaney P, Ouzman S (2016) Plants before farming: The deep history of plant-use and representation in the rock art of Australia's Kimberley region. *Quaternary International* 489: 26-45

2. Crawford IM (1982) Traditional Aboriginal plant resources in the Kalumburu area: aspects in ethno-economics. *Records of the Western Australian Museum* Suppl. 15

3. Karadada J, Karadada L, Goonak W, Mangolamara G, Bunjuck W, Karadada L, Djanghara B, Mangolamara S, Oobagooma J, Charles A, Williams D, Karadada R, Saunders T, Wightman G (2011) Uunguu plants and animals: Aboriginal biological knowledge from Wunambal Gaambera country in the north-west Kimberley, Australia. Northern Territory Botanical Bulletin no. 35. Wunambal Gaambera Aboriginal Corporation, Wyndham, Australia

4. Vigilante T, Toohey J, Gorring A, Blundell V, Saunders T, Mangolamara S, George K, Oobagooma J, Waina M, Morgan K, Doohan K (2013) Island country: Aboriginal connections, values and knowledge of the Western Australian Kimberley islands in the context of an island biological survey. *Records of the Western Australian Museum* Suppl 81: 145-182

5. Mangolomara S, Karadada L, Oobagooma J, Woolagoodja D, Karadada J, Doohan K (2018) *Nyara pari kala niragu (Gaambera), gadawara ngyaran-gada (Wunambal), inganinja gubadjoongana (Woddordda) = we are coming to see you.* Dambimangari Aboriginal Corporation and Wunambal Gaambera Aboriginal Corporation, Derby, Australia

6. Hill R, Pert PL, Davies J, Robinson CJ, Walsh F, Falco-Mammone F (2013) Indigenous land management in Australia: Extent, scope, diversity, barriers and success factors. CSIRO, Cairns, Australia

7. Mangolamara G, Burbidge AA, Fuller PJ (1991) Wunumbal words for rainforest and other Kimberley plants and animals. In: McKenzie NL, Johnston RB, Kendrick PG (eds) *Kimberley rainforests of Australia*, pp 413-421. Surrey Beatty & Sons, Chipping Norton, Australia

8. Belfield E, Beames L, Black S (2012) Valuable & threatened: monsoon vine thickets of the Dampier Peninsula. A summary of key findings from the Broome Botanical Society. Environs Kimberley, Broome, Australia

9. Black SJ, Willing T, Dureau DM (2010) A comprehensive survey of the flora, extent and condition of vine thickets on coastal sand dunes of Dampier Peninsula, West Kimberley 2000-2002. Broome Botanical Society, Broome, Australia

10. Beames L (2013) Valuable & endangered: Working together to understand and manage threats to monsoon vine thickets of the Dampier Peninsula. West Kimberley Nature Project 2011-2013. Environs Kimberley, Broome, Australia

11. Vigilante T, Ondei S, Goonack C, Williams D, Young P, Bowman DMJS (2017) Collaborative research on the ecology and management of the 'Wulo' monsoon rainforest in Wunambal Gaambera Country, North Kimberley, Australia. *Land* 6(4): 68

CHAPTER 4: DOCUMENTING KIMBERLEY RAINFORESTS

1. Grant-Richards E (1906) *Dampier's voyages by Captain William Dampier.* Volume 1. E Grant-Richards, London, UK

2. George AS (1999) *William Dampier in New Holland: Australia's first natural historian.* Blooming Books, Hawthorn, Australia

3. King PP (1827) *Narrative of a survey of the intertropical and western coast performed between the years 1818 and 1822*. Volumes 1 and 2. John Murray, London, UK

4. Orchard AE, Orchard TA (2018) The Australian botanical journals of Allan Cunningham: the later King expeditions, February 1819 - September 1822. Orchard AE and Orchard TA, Weston Creek, Australia

5. Brockman JG (1880) Journal of an exploring trip from Beagle Bay to the Fitzroy River and back again. *The West Australian*, 28 May

6. Johnstone RE (1981) Notes on the distribution, ecology and taxonomy of the Red-crowned Pigeon (Ptilinopus regina) and Torres Strait Pigeon (*Ducula bicolor*) in Western Australia. *Records of the Western Australian Museum* 9(1): 7-22

7. Hill G (1911) Field notes on the birds of the Kimberley, north-west Australia. *Emu* 10: 258-290

8. Gardner CA (1923) Botanical notes: Kimberley Division of Western Australia. Western Australia Forests Department, Bulletin no. 32. Government of Western Australia, Perth, Australia

9. Easton WR (1923) Report on the North Kimberley district of Western Australia. Department of the North-West, Publication no. 3. Government of Western Australia, Perth, Australia

10. Gardner CA (1944) The vegetation of Western Australia. Presidential Address, 1942. *Journal of the Royal Society of Western Australia* 28: xi-lxxxviii

11. Hooker JD (1859) On the flora of Australia, its origin, affinities, and distribution; being an introductory essay to the Flora of Tasmania. In: *Flora Tasmaniae*, Volume 1, pp i-cxxviii. Lovell Reeve, London, UK

12. Christian CS, Stewart GA (1953) General report on survey of Katherine-Darwin region, 1946. Land Research Series no. 1. CSIRO, Melbourne, Australia

13. Speck NH (1960) Vegetation of the North Kimberley area, W.A. In: The lands and pastoral resources of the North Kimberley area, W.A., pp 41-63. Land Research Series no. 4. CSIRO, Melbourne, Australia

14. Allen AD (1966) Geology of the Montague Sound 1:250,000 area (SD/51-12), Western Australia. Australian Record 1966/20. Bureau of Mineral Resources, Canberra, Australia

15. Allen AD (1971) Explanatory notes on the Montague Sound geological sheet area. Bureau of Mineral Resources, Canberra, Australia

16. Beard JS (1976) The monsoon forests of the Admirality Gulf, Western Australia. *Vegetatio* 31: 177-192

17. Burbidge A, McKenzie NL (1978) The islands of the north-west Kimberley, Western Australia. *Wildlife Research Bulletin Western Australia* 7: 1-47

18. Miles JM, Burbidge AA (eds) (1975) *A biological survey of the Prince Regent River Reserve, north-west Kimberley, Western Australia, in August 1974*. Department of Fisheries and Wildlife, Perth, Australia

19. Kabay ED, Burbidge AA (eds) (1977) *A biological survey of the Drysdale National Park, North Kimberley, Western Australia in August 1975*. Government of Western Australia, Perth, Australia

20. Hnatiuk RJ, Kenneally KF (1981) A survey of the vegetation and flora of Mitchell Plateau, Kimberley, Western Australia. In: *Biological survey of Mitchell Plateau and Admiralty Gulf, Kimberley, Western Australia*. Western Australian Museum, Perth, Australia

21. Beard JS, Clayton-Greene KA, Kenneally KF (1984) Notes on the vegetation of the Bougainville Peninsula, Osborn and Institut Islands, Northern Kimberley District, Western Australia. *Vegetatio* 57: 3-13

22. Clayton-Green KA, Beard JS (1985) The fire factor in vine thicket and woodland vegetation of the Admiralty Gulf region, north-west Kimberley, Western Australia. In: Ridpath MG, Corbett LK (eds) *Ecology of the wet-dry tropics*, pp 225-230. Ecological Society of Australia, Melbourne, Australia

23. Beard JS, Kenneally KF (1993) Dry coastal ecosystems of Northern Australia. In: Van der Maarel E (ed) *Ecosystems of the world, 2B: Dry coastal ecosystems: Africa, America, Asia and Oceania*, pp 239-258. Elsevier, Amsterdam, Netherlands

24. McKenzie NL, Johnston RB, Kendrick PG (eds) (1991) *Kimberley rainforests of Australia*. Surrey Beatty & Sons, Chipping Norton, Australia

25. Comrie-Greig J, Abdo L (eds) (2014) Ecological studies of the Bonaparte Archipelago and Browse Basin. INPEX Operations Australia, Perth, Australia

26. How R, Schmitt L, Teale R, Cowan M (2006) Appraising vertebrate diversity on Bonaparte Islands, Kimberley, Western Australia. *The Western Australian Naturalist* 25(2): 92-110

27. How RA, Spencer PBS, Schmitt LH (2009) Island populations have high conservation value for northern Australia's top marsupial predator ahead of a threatening process. *Journal of Zoology* 278(3): 206-217

28. Gibson LA, Yates S, Doughty P (eds) (2014) Biodiversity values on selected Kimberley Islands, Australia. *Records of the Western Australian Museum* Suppl. 81

CHAPTER 5: CLIMATE AND WEATHER IN THE KIMBERLEY

1. Keenan TD, Morton BR, Manton MJ, Holland GJ (1989) The Island thunderstorm experiment (ITEX) - a study of tropical thunderstorms in the maritime continent. *Bulletin of the American Meteorological Society* 7: 152-159

2. Keenan T, Rutledge S, Carbone R, Wilson J, Takahashi T, May P, Tapper N, Platt M, Hacker J, Sekelsky S, Moncrieff M, Saito K, Holland G, Crook A, Gage K (2000) The Maritime Continent Thunderstorm Experiment (MCTEX): overview and some results. *Bulletin of the American Meteorological Society* 81(10): 2433-2456

3. Suppiah R (1992) The Australian summer monsoon: a review. *Progress in Physical Geography: Earth and Environment* 16(3): 283-318

4. Beard JS (1979) *Vegetation survey of Western Australia, sheet 1: Kimberley*. University of Western Australia Press, Nedlands, Australia

5. Department of Water and Environmental Regulation (2017) Groundwater-dependent ecosystems of the Dampier Peninsula. Environmental Water Report Series, Report no. 29. Government of Western Australia, Perth, Australia

6. Collins B (2012) The mysteries and magic of fog in Broome. ABC, 24 August

7. Van Damme P (1991) Plant ecology of the Namib Desert. *Afrika Focus* 7(4): 355-400

8. Shanyengana ES, Henschel JR, Seely MK, Sanderson RD (2002) Exploring fog as a supplementary water source in Namibia. *Atmospheric Research* 64(1-4): 251-259

9. Kimberley Foundation Australia (2014) The changing climate of the Kimberley. *Kimberley Foundation Australia Newsletter* Autumn: 6. Kimberley Foundation Australia, Melbourne, Australia

10. Denniston RF, Villarini G, Gonzalez AN, Wyrwoll K-H, Polyak VJ, Ummenhofer CC, Lachniet MS, Wanamaker Jr AD, Humphreys WF, Woods D, Cugley J (2015) Extreme rainfall activity in the Australian tropics reflects changes in the El Nino/Southern Oscillation over the last two millennia. *Proceedings of the National Academy of Sciences* USA 112(15): 4576-4581

11. Quicke A (2019) HeatWatch: Extreme heat in the Kimberley. The Australia Institute, Canberra, Australia

12. Kenneally KF, Edinger DC, Willing T (1996) *Broome and beyond: Plants and people of the Dampier Peninsula, Kimberley, Western Australia*. Western Australian Department of Conservation and Land Management, Perth, Australia

13. Karadada J, Karadada L, Goonak W, Mangolamara G, Bunjuck W, Karadada L, Djanghara B, Mangolamara S, Oobagooma J, Charles A, Williams D, Karadada R, Saunders T, Wightman G (2011) Uunguu plants and animals: Aboriginal biological knowledge from Wunambal Gaambera country in the north-west Kimberley, Australia. Northern Territory Botanical Bulletin no. 35. Wunambal Gaambera Aboriginal Corporation, Wyndham, Australia

14. Crawford IM (1968) *The art of the Wandjina*. Oxford University Press, Melbourne, Australia

15. O'Connor SJ, Balme J, Fyfe J, Oscar J, Oscar M, Davis J, Malo H, Nugget R, Surprise D (2013) Marking resistance? Change and continuity in the recent rock art of the southern Kimberley, Australia. *Antiquity* 87(336): 539-554

CHAPTER 6: MONSOON RAINFOREST ENVIRONMENTS

1. Donaldson M (2012) *Kimberley rock art. Volume 1: Mitchell Plateau area*. Wildrock Publications, Perth, Australia

2. Tyler IM, Sheppard S, Pirajno F, Griffin TJ (2006) Hart-Carson LIP, Kimberley region, northern Western Australia. August 2006 LIP of the month, Large Igneous Provinces Commission

3. Stoneman TC, McArthur WM, Walsh FJ (1991) Soils and landforms of Kimberley rainforests, Western Australia. In: McKenzie NL, Johnston RB, Kendrick PG (eds) *Kimberley rainforests of Australia*, pp 247-265. Surrey Beatty & Sons, Chipping Norton, Australia

4. Allen AD (1971) Explanatory notes on the Montague Sound geological sheet area. Bureau of Mineral Resources, Canberra, Australia

5. Beard JS (1976) The monsoon forests of the Admirality Gulf, Western Australia. *Vegetatio* 31: 177-192

6. Beard JS, Kenneally KF (1993) Dry coastal ecosystems of Northern Australia. In: Van der Maarel E (ed) *Ecosystems of the world, 2B: Dry coastal ecosystems: Africa, America, Asia and Oceania*, pp 239-258. Elsevier, Amsterdam, Netherlands

CHAPTER 7: THE BIOCOMPLEXITY OF MONSOON RAINFORESTS

1. McKenzie NL, Johnston RB, Kendrick PG (eds) (1991) *Kimberley rainforests of Australia*. Surrey Beatty & Sons, Chipping Norton, Australia

2. Kenneally KF, McKenzie NL (1991) *Companion to Kimberley rainforests of Australia*. Surrey Beatty & Sons: Chipping Norton, Australia

3. Whitmore TC (1984) *Tropical rainforests of the Far East*. Clarendon Press, Oxford, UK

4. Giesen W, Wulffraat S, Zieren M, Scholten L (2006) Mangrove guidebook for Southeast Asia. Food and Agriculture Organization of the United Nations and Wetlands International, Bangkok, Thailand

5. Joyce EM, Thiele KR, Ferry Slik JW, Crayn DM (2021) Plants will cross the lines: climate and available land mass are the major determinants of phytogeographical patterns in the Sunda-Sahul Convergence Zone. *Biological Journal of the Linnean Society* 132(2): 374-387

6. Sniderman JMK, Jordan GJ (2011) Extent and timing of floristic exchange between Australian and Asian rain forests. *Journal of Biogeography* 38(8): 1445-1455

7. Webb L, Tracey JG (1981) Australian rainforests: patterns and change. In: Keast A (ed) *Ecological biogeography of Australia*, pp 605-694. Dr W Junk, The Hague, The Netherlands

8. Wegener A (1966) *The origin of continents and oceans*. Dover Publications, New York, NY, USA

9. Truswell EM (1993) Vegetation changes in the Australian Tertiary in response to climatic and phytogeographic forcing factors. *Australian Systematic Botany* 6(6): 533-557

10. Macphail MK (2007) Australian palaeoclimates: cretaceous to tertiary: a review of palaeobotanical and related evidence to the year 2000. Cooperative Research Centre for Landscape Environments and Mineral Exploitation, Bentley, Australia

11. Murphy DJ, Crayn DM (2017) Australian comparative phytogeography: a review. In: Ebach M (ed) *Handbook of Australasian biogeography*, pp 129-154. CRC Press, Boca Raton, FL, USA

12. McLoughlin S (2001) The breakup history of Gondwana and its impact on pre-Cenozoic floristic provincialism. *Australian Journal of Botany* 49(3): 271-300

13. Byrne M, Yeates DK, Joseph L, Kearney M, Bowler J, Williams MAJ, Cooper S, Donnellan SC, Keogh JS, Leys R, Melville J, Murphy DJ, Porch N, Wyrwoll K-H (2008) Birth of a biome: insights into the assembly and maintenance of the Australian arid zone biota. *Molecular Ecology* 17(20): 4398-4417

14. Kershaw AP (1988) The late Cainozoic history of Australasia: 20 million to 20 thousand years BP. In: Huntly B, Webb T (eds) *Vegetation history*, pp 237-301. Kluwer, Dordrecht, The Netherlands

15. Barlow BA, Hyland BPM (1988) The origins of the flora of Australia's wet tropics. In: R. Kitching R (ed) *The ecology of Australia's wet tropics: proceedings of a symposium held at the University of Queensland, Brisbane, August 25-27, 1986*. Surrey Beatty & Sons, Chipping Norton, Australia

16. Joyce EM, Pannell CM, Rossetto M, Yap J-YS, Thiele KR, Wilson PD, Crayn DM (2021) Molecular phylogeography reveals two geographically and temporally separated floristic exchange tracks between Southeast Asia and northern Australia. *Journal of Biogeography* 48: 1213-1227

17. Nix H (1982) Environmental determinants of biogeography and evolution in Terra Australis. In: Barker WR, Greenslade PJM (eds) *Evolution of the flora and fauna of arid Australia*, pp 47-66. Peacock Press, Frewville, Australia

18. Truswell EM, Harris WK (1982) The Cainozoic palaeobotanical record in arid Australia: fossil evidence for the origin of an arid adapted flora. In: Barker WR, Greenslade PJM (eds) *Evolution of the fauna and flora of arid Australia*, pp 67-76. Peacock Publications, Frewville, Australia

19. Floyd AG (1990) *Australian rainforests in New South Wales*. Surrey Beatty & Sons, Chipping Norton, Australia

20. Crisp M, Cook L, Steane D (2004) Radiation of the Australian flora: what can comparisons of molecular phylogenies across multiple taxa tell us about the evolution of diversity in present-day communities? *Philosophical Transactions of the Royal Society B: Biological Sciences* 359(1450): 1551-1571

21. Chappell J, Thom BG (1977) Sea levels and coasts. In: Allen J, Godson J, Jones R (eds) *Sunda and Sahul: prehistoric studies in south-east Asia, Melanesia and Australia*, pp 275-291. Academic Press, London, UK

22. Jennings JN (1975) Desert dunes and estuarine fill in the Fitzroy estuary (North-Western Australia). *Catena* 2: 215-262

23. Thom BG, Wright LD, Coleman JM (1975) Mangrove ecology and deltaic-estuarine geomorphology: Cambridge Gulf-Ord River, Western Australia. *Journal of Ecology* 63(1): 203-232

24. Proske U, Heslop D, Haberle S (2014) A Holocene record of coastal landscape dynamics in the eastern Kimberley region, Australia. *Journal of Quaternary Science* 29(2): 163-174

25. Field E, McGowan HA, Moss PT, Marx SK (2017) A late Quaternary record of monsoon variability in the northwest Kimberley, Australia. *Quaternary International* 449: 119-135

26. Ishiwa T, Yokoyama Y, Reuning L, McHugh CM, De Vleeschouwer D, Gallagher SJ (2019) Australian Summer Monsoon variability in the past 14,000 years revealed by IODP Expedition 356 sediments. *Progress in Earth and Planetary Science* 6: 17

27. Denniston RF, Wyrwoll K-H, Polyak VJ, Brown JR, Asmerom Y, Wanamaker Jr AD, LaPointe Z, Ellerbroek R, Barthelmes M, Cleary D, Cugley J, Woods D, Humphreys WF (2013) A Stalagmite record of Holocene Indonesian-Australian summer monsoon variability from the Australian tropics. *Quaternary Science Reviews* 78: 155-168

28. Denniston RF, Villarini G, Gonzalez AN, Wyrwoll K-H, Polyak VJ, Ummenhofer CC, Lachniet MS, Wanamaker Jr AD, Humphreys WF, Woods D, Cugley J (2015) Extreme rainfall activity in the Australian tropics reflects changes in the El Nino/Southern Oscillation over the last two millennia. *Proceedings of the National Academy of Sciences USA* 112(15): 4576-4581

29. May TW (2001) Documenting the fungal biodiversity of Australasia: from 1800 to 2000 and beyond. *Australian Systematic Botany* 14(3): 329-356

30. Henson M, Kenneally K, Griffin EA, Barrett RL (2014) Terrestrial flora. In: Comrie-Greig J, Abdo L (eds) Ecological studies of the Bonaparte Archipelago and Browse Basin, pp 19-102. INPEX Operations Australia, Perth, Australia

31. Dixon KW (2002) Orchids of Kimberley. *Boab Bulletin* 48: 4-5

32. Barrett RL (2015) Fifty new species of vascular plants from Western Australia - celebrating fifty years of the Western Australian Botanic Garden at Kings Park. *Nuytsia* 26: 3-20

33. Warren JM, Emamdie DZ, Kalai (1997) Reproductive allocation and pollinator distribution in cauliflorous trees in Trinidad. *Journal of Tropical Ecology* 13(3): 337-345

34. Kinnaird M (2000) Big on figs. *National Wildlife* 30(1): 12-22

35. Shanahan M, So S, Compton SG, Corlett R (2001) Fig-eating by vertebrate frugivores: a global review. *Biological Reviews of the Cambridge Philosophical Society* 76(4): 529-572

36. Clarkson JR, Kenneally KF (1988) The floras of Cape York and the Kimberley: a preliminary comparative analysis. *Proceedings of the Ecological Society of Australia* 15: 259-266

37. Kenneally KF, Keighery GJ, Hyland BPM (1991) Floristics and phytogeography of Kimberley rainforests, Western Australia. In: McKenzie NL, Johnston RB, Kendrick PG (eds) *Kimberley rainforests of Australia*, pp 93-131. Surrey Beatty & Sons, Chipping Norton, Australia

38. Pannell CM (2008) A key to Aglaia (Meliaceae) in Australia, with a description of a new species, A. cooperae, from Cape York Peninsula, Queensland. *Journal of the Adelaide Botanic Gardens* 22: 67-71

39. Hyland BPM, Whiffin T, Cristophel DC, Gray B, Elick RW (2003) *Australian tropical rain forest plants: trees, shrubs and vines*. CSIRO Publishing, Collingwood, Australia

40. Fryxell PA (1987) Three new species (from Australia and Venezuela) and three new names (of Mexican plants) in the Malvaceae. *Systematic Botany* 12(2): 274-280

41. Harrington MG, Jackes BR, Barrett MD, Craven LA, Barrett RL (2012) Phylogenetic revision of Backhousieae (Myrtaceae): Neogene divergence, a revised circumscription of *Backhousia* and two new species. *Australian Systematic Botany* 25(6): 404-417

42. Kenneally KF (1983) Kalanchoe crenata (Andr.) Haw (Crassulaceae) a new record for Australia. *Australasian Systematic Botany Society Newsletter* 34: 4-5

43. Ohba H (2003) Taxonomic studies on the Asian species of the genus *Kalanchoe* (Crassulaceae) 1. *Kalanchoe spathulata* and its allied species. *Journal of Japanese Botany* 78(5): 247-256

44. Groombridge B (ed) (1992) *Global biodiversity: status of the Earth's living resources*. Chapman & Hall, London, UK

45. Wilson BR (1981) General introduction. In: *Biological Survey of Mitchell Plateau and Admiralty Gulf, Kimberley, Western Australia*. Western Australian Museum, Perth, Australia

46. Solem A (1984) A world model of land snail diversity and abundance. In: Solem A, Van Bruggen AC (eds) *World-wide snails: Biogeographical studies on non-marine mollusca*, pp 6-22. EJ Brill, Leiden, The Netherlands

47. Solem A (1984) Caemaenid land snails from Western and Central Australia (Mollusca: Pulmonata: Camaenidae), IV. Taxa from the Kimberley, Westraltrachia Iredale, 1933 and related genera. *Records of the Western Australian Museum* Suppl 17: 426-705

48. Cameron RAD, Pokryszko BM, Wells FE (2005) Alan Solem's work on the diversity of Australasian land snails: an unfinished project of global significance. *Records of the Western Australian Museum* Suppl 68: 1-10

49. Solem A (1991) Land snails of Kimberley rainforest patches and biogeography of all Kimberley land snails. In: McKenzie NL, Johnston RB, Kendrick PG (eds) *Kimberley rainforests of Australia*, pp 145-246. Surrey Beatty & Sons, Chipping Norton, Australia

50. Solem A, McKenzie NL (1991) The composition of land snail assemblages in Kimberley rainforests. In: McKenzie NL, Johnston RB, Kendrick PG (eds) *Kimberley rainforests of Australia*, pp 247-265. Surrey Beatty & Sons, Chipping Norton, Australia

51. Gibson LA, Yates S, Doughty P (eds) (2014) Biodiversity values on selected Kimberley Islands, Australia. *Records of the Western Australian Museum* Suppl. 81

52. Criscione F, Köhler F (2014) Molecular phylogenetics and comparative anatomy of *Kimberleytrachia* Köhler, 2011 - a genus of land snails endemic to the coastal Kimberley, Western Australia with description of new taxa (Gastropoda, Camaenidae). *Contributions to Zoology* 83(4): 245-267

53. Harvey MS (2014) Arachnida (Arthropoda: Chelicerata) of Western Australia: overview and prospects. *Journal of the Royal Society of Western Australia* 97: 57-64

54. Köhler F, Criscione F (2013) Plio-Pleistocene out-of-Australia dispersal in a camaenid land snail. *Journal of Biogeography* 40(10): 1971-1982

55. Fisher J, Beames L, Bardi Jawi Rangers, Nyul Nyul Rangers, Majer J, Heterick B (2014) Using ants to monitor changes within and surrounding the endangered monsoon vine thickets of the tropical Dampier Peninsula, north Western Australia. *Forest Ecology and Management* 318: 78-90

56. Naumann ID, Weir TA, Edwards ED (1991) Insects of Kimberley rainforests. In: McKenzie NL, Johnston RB, Kendrick PG (eds) *Kimberley rainforests of Australia*, pp 299-332. Surrey Beatty & Sons, Chipping Norton, Australia

57. Price OF, Woinarski JCZ, Robinson D (1999) Very large area requirements for frugivorous birds in monsoon rainforests of the Northern Territory, Australia. *Biological Conservation* 91(2-3): 169-180

CHAPTER 8: RAINFOREST ON SAND DUNES

1. Beames L (2013) Valuable & endangered: Working together to understand and manage threats to monsoon vine thickets of the Dampier Peninsula. West Kimberley Nature Project 2011-2013. Environs Kimberley, Broome, Australia

2. McKenzie NL, Kenneally KF (1983) Wildlife of the Dampier Peninsula, Part 1: Background and environment. *Wildlife Research Bulletin of Western Australia* 11: 5-23

3. Kenneally KF, Edinger DC, Willing T (1996) *Broome and beyond: Plants and people of the Dampier Peninsula, Kimberley, Western Australia.* Western Australian Department of Conservation and Land Management, Perth, Australia

4. McKenzie NL, Johnston RB, Kendrick PG (eds) (1991) *Kimberley rainforests of Australia*. Surrey Beatty & Sons, Chipping Norton, Australia

5. Black SJ, Willing T, Dureau DM (2010) A comprehensive survey of the flora, extent and condition of vine thickets on coastal sand dunes of Dampier Peninsula, West Kimberley 2000-2002. Broome Botanical Society, Broome, Australia

CHAPTER 9: MITCHELL PLATEAU: A FOCUS AREA

1. Bradley AJ, Kemper CM, Kitchener DJ, Humphreys WF, How RA (1987) Small mammals of the Mitchell Plateau Region, Kimberley, Western Australia. *Australian Wildlife Research* 14(4): 397-413

2. Kenneally KF, Edinger D, Coate K, Hyland B, How R, Schmitt L, Cowan M, Willing T, Done C (2003) The Last Great Wilderness - Exploring the Mitchell Plateau 2002. Landscope Expeditions Report no. 49. Western Australian Department of Conservation and Land Management, Perth, Australia

3. Western Australian Museum (1981) *Biological survey of Mitchell Plateau and Admiralty Gulf, Kimberley, Western Australia*. Western Australian Museum, Perth, Australia

4. Burbidge AA, McKenzie NL, Brennan KEC, Woinarski JCZ, Dickman CR, Baynes A, Gordon G, Menkhorst PW, Robinson AC (2008) Conservation status and biogeography of Australia's terrestrial mammals. *Australian Journal of Zoology* 56: 411-422

5. Gibson L, McKenzie NL (2012) Identification of biodiversity assets on selected Kimberley Islands: background and implementation. *Records of the Western Australian Museum* 81(Suppl): 1-14

6. Russell-Smith J, McKenzie NL, Woinarski JCZ (1992) Conserving vulnerable habitat in northern and north-western Australia: the rainforest archipelago. In: Moffatt I, Webb A (eds) *Conservation and development issues in Northern Australia*, pp 63-68. North Australia Research Unit, Australian National University, Darwin, Australia

7. Bowman DMJS, Woinarski JCZ (1994) Biogeography of Australian monsoon rainforest mammals: implications for the conservation of rainforest mammals. *Pacific Conservation Biology* 1(2): 98-106

8. Kendrick PG, Rolfe JK (1991) The reptiles and amphibians of Kimberley rainforest. In: McKenzie NL, Johnston RB, Kendrick PG (eds) *Kimberley rainforests of Australia*, pp 347-359. Surrey Beatty & Sons, Chipping Norton, Australia

CHAPTER 10: ETHICAL AND SUSTAINABLE USE OF THE KIMBERLEY'S BIOLOGICAL RESOURCES

1. Ens EJ, Pert P, Clarke PA, Budden M, Clubb L, Doran B, Douras C, Gaikwad J, Gott B, Leonard S, Locke J, Packer J, Turpin G, Wason S (2015) Indigenous biocultural knowledge in ecosystem science and management: Review and insight from Australia. *Biological Conservation* 181: 133-149

2. Ouzman S, Veth P, Myers C, Heaney P, Kenneally K (2017) Plants before animals? Aboriginal rock art as evidence of ecoscaping in Australia's Kimberley. In: David B, McNiven I (eds) *The Oxford handbook of the archaeology and anthropology of rock art*, pp 469-480. Oxford University Press, Oxford, UK

3. Kenneally KF, Edinger DC, Willing T (1996) *Broome and beyond: Plants and people of the Dampier Peninsula, Kimberley, Western Australia*. Western Australian Department of Conservation and Land Management, Perth, Australia

4. Jones N (2016) Kimberley seed bank: Traditional knowledge used to protect biodiversity. *ABC News*, 27 March

5. Lindsay M, Beames L, Yawuru Country Managers, Nyul Nyul Rangers, Bardi Jawi Rangers (2022) Integrating scientific and Aboriginal knowledge, practice and priorities to conserve an endangered rainforest ecosystem in the Kimberley region, northern Australia. *Ecological Management and Restoration* 23(S1): 93-104

6. Lethlean J (2017) Best of the bush. *The Australian*, 14 February

7. Hegarty MP, Hegarty EE, Wills RBH (2001) Food safety of Australian plant bushfoods. RIRDC publication 01/28. Rural Industries Research and Development Corporation, Barton, Australia

8. Packer J, Turpin G, Ens E, Venkataya B, Mbabaram Community, Yirralka Rangers, Hunter J (2019) Building partnerships for linking biomedical science with traditional knowledge of customary medicines: a case study with two Australian communities. *Journal of Ethnobiology and Ethnomedicine* 15: 69

9. Stoutjesdijk P (2013) Plant genetic resources for food and agriculture: second national report - Australia. Australian Government, Canberra, Australia

10. Medhi K, Deka M, Bhau BS (2013) The genus *Zanthoxylum* - A stockpile of biological and ethnomedicinal properties. *Open Access Scientific Reports* 2: 3

11. Department of Health and Aged Care (2001) *Aristolochia* fact sheet. Australian Government, Canberra, Australia

12. Natarajan K, Narayanan N, Ravichandran N (2012) Review on "Mucuna" - the wonder plant. *International Journal of Pharmaceutical Sciences Review and Research* 17(1): 86-93

13. Powell R, Murdoch L (2010) Patent erupts over Kakadu plum. *The Sydney Morning Herald*, 4 December

14. Robinson D, Raven M (2016) Identifying and preventing biopiracy in Australia: patent landscapes and legal geographies for plants with Indigenous Australian uses. *Australian Geographer* 48(3): 311-331

15. Mills V (2015) Favourite Kimberley bush foods and medicine targeted by bio-pirate. *ABC*, 5 February

16. Gan D, Hines M, Aravena J, Jones B (2020) Compositions comprising kakadu plum extract or acai berry extract (US patent no. 2013/0149401A1). US Patent and Trademark Office, Alexandria, VA, USA

17. Convention on Biological Diversity (1993) Article 8(j): Traditional knowledge, innovations and practices. Convention on Biological Diversity, Montreal, Canada

18. Blakeney M (2017) Bioprospecting and Traditional knowledge in Australia. In: McManis CR, Ong B (eds) *Routledge handbook of biodiversity and the law*, pp 254-275. Routledge, Abingdon, UK

19. Environmental Defenders Office (2016) White Paper: Western Australia's Biodiversity Conservation Bill 2015: Review and recommendations. Environmental Defenders Office, Perth, Australia

20. House of Representatives Standing Committee on Primary Industries and Regional Services (2001) Bioprospecting: Discoveries changing the future. Parliament of the Commonwealth of Australia, Canberra, Australia

21. Linge K, Patterson A (2019) Report of bioprospecting workshop convened by Chief Scientist of WA. ChemCentre ref. 1854644. Government of Western Australia, Bentley, Australia

22. Davis M (1998) Biological diversity and Indigenous knowledge. Research Paper 17. Parliament of Australia, Canberra, Australia

CHAPTER 11: MANAGING THREATS TO KIMBERLEY RAINFORESTS

1. Tyler IM, Hocking RM, Haines PW (2012) Geological evolution of the Kimberley region of Western Australia. Episodes: *Journal of International Geoscience* 35(1): 298-306

2. Kimberley Development Commission (2020) Kimberley economy snapshot. Government of Western Australia, Perth, Australia

3. Kimberley Development Commission (2021) 2020-21 Annual Report. Government of Western Australia, Perth, Australia

4. Department of Fisheries (2013) Kimberley aquaculture development zone project. Government of Western Australia, Perth, Australia

5. Carwardine J, O'Connor T, Legge S, Mackey B, Possingham HP, Martin TG (2011) Priority threat management to protect Kimberley wildlife. CSIRO Ecosystem Sciences, Brisbane, Australia

6. Jackson W (2017) Five-yearly environmental stocktake highlights the conflict between economy and nature. *The Conversation*, 7 March

7. Regional Development Australia Kimberley (2020) Business in the Kimberley. Regional Development Australia Kimberley, Broome, Australia

8. Scherrer P, Smith A, Dowling R (2008) Tourism and the Kimberley coast waterways: Environmental and cultural aspects of expedition cruising. Cooperative Research Centre for Sustainable Tourism, Gold Coast, Australia

9. Tourism Research Services (2015) The economic benefits of creating a 'world class' Great Kimberley Marine Park. Murdoch University, Perth, Australia

10. Semeniuk V, Kenneally KF, Wilson PG (1978) Mangroves of Western Australia. Handbook no. 12. WA Naturalists' Club, Perth, Australia

11. Sucharitakul G, Hardy J (2021) A threat and a solution - tourism's role in mangrove protection. *Race to Zero* website, 26 July

12. McKenzie NL, Start AN, Burbidge AA, Kenneally KF, Burrows ND (2009) Protecting the Kimberley: a synthesis of scientific knowledge to support conservation management in the Kimberley region of Western Australia. Part B: Terrestrial environments. Department of Environment and Conservation, Perth, Australia

13. Bradstock RA (2008) Effects of large fires on biodiversity in south-eastern Australia: disaster or template for diversity? *International Journal of Wildland Fire* 17(6): 809-822

14. Keane RE, Agee JK, Fule P, Keeley JE, Key C, Kitchen SG, Miller R, Schulte LA (2008) Ecological effects of large fires on US landscapes: benefit or catastrophe? *International Journal of Wildland* Fire 17: 696-712

15. Russell-Smith, J, Cook GD, Cooke PM, Edwards AC, Lendrum M, Meyer CP, Whitehead PJ (2013) Managing fire regimes in north Australian savannas: applying Aboriginal approaches to contemporary global problems. *Frontiers in Ecology and the Environment* 11(s1): e55-e63

16. Vigilante T, Bowman DMJS (2015) Effects of fire history and the structure and floristic composition of woody vegetation around Kalumburu, north Kimberley, Australia: a landscape-scale natural experiment. *Australian Journal of Botany* 52: 381-404

17. Keighery GJ (2014) Protecting our Kimberley's unique flora. *Landscope* 29(4): 9-13

18. Ondei S, Prior LD, Vigilante T, Bowman DMJS (2017) Fire and cattle disturbance affects vegetation structure and rain forest expansion into savanna in the Australian monsoon tropics. *Journal of Biogeography* 44(10): 2331-2342

19. Banfai DS, Bowman DMJS (2007) Drivers of rain-forest boundary dynamics in Kakadu National Park, northern Australia: A field assessment. *Journal of Tropical Ecology* 23(1): 73-86

20. Ondei S, Prior LD, Williamson GJ, Vigilante T, Bowman DMJS (2017) Water, land, fire, and forest: Multi-scale determinants of rainforests in the Australian monsoon tropics. *Ecology & Evolution* 7(5): 1592-1604

21. Ondei S, Prior LD, Vigilante T, Bowman DMJS (2016) Post-fire resprouting strategies of rainforest and savanna saplings along the rainforest-savanna boundary in the Australian monsoon tropics. *Plant Ecology* 217: 711-724

22. Vogel E, Meyer C, Eckard R (2017) Severe heatwaves show the need to adapt livestock management for climate. *The Conversation*, 28 February

23. Corey B, Radford I, Carnes K, Moncrieff A (2016) North-Kimberley Landscape Conservation Initiative: 2013-14 monitoring, evaluation, research & improvement report. Department of Parks and Wildlife, Kununurra, Australia

24. Russell-Smith J, Murphy BP, Meyer CP, Cook GD, Maier S, Edwards AC, Schatz J, Brocklehurst P (2009) Improving estimates of savanna burning emissions for greenhouse accounting in northern Australia: limitations, challenges, applications. *International Journal of Wildland Fire* 18(1): 1-18

25. Clean Energy Regulator (2015) Participating in the Emissions Reduction Fund: a guide to the savanna fire management method 2015. Australian Government, Canberra, Australia

26. Radford IJ, Gibson LA, Corey B, Carnes K, Fairman R (2015) Influence of fire mosaics, habitat characteristics and cattle disturbance on mammals in fire-prone savanna landscapes of the northern Kimberley. *PLoS ONE* 10(6): e0130721

27. Radford I, Thomson-Dans C, Fairman R, Hatherley E (2013) Kimberley mammals bouncing back. *Landscope* 28(3): 32-38

28. Department of Parks and Wildlife (2014) Cane toad strategy for Western Australia 2014-2019. Government of Western Australia, Perth, Australia

29. Keighery GJ (2015) Weeding out Kimberley weeds. *Landscope* 31(1): 41-45

30. Vilà M, Espinar JL, Hejda M, Hulme PE, Jarošík V, Maron JL, Pergl J, Schaffner U, Sun Y, Pyšek P (2011) Ecological impacts of invasive alien plants: a meta-analysis of their effects on species, communities and ecosystems. *Ecology Letters* 14(7): 702-708

31. Oerke E-C (2006) Crop losses to pests. *Journal of Agricultural Science* 144(1): 31–43

32. Read I (2015) Weed of the week – Siratro (*Macroptilium atropurpureum*). The Courier Mail, 5 September

33. Webber BL, Yeoh PB, Scott JK (2014) Invasive *Passiflora foetida* in the Kimberley and Pilbara: understanding the threat and exploring solutions. CSIRO, Perth, Australia

34. Hopley T, Webber BL, Raghu S, Morin L, Byrne M (2021) Revealing the introduction history and phylogenetic relationships of *Passiflora foetida sensu lato* in Australia. *Frontiers in Plant Science* 12: 651805

35. Department of Environment and Science (2021) A field guide to assessing Australia's tropical riparian zones through the National Action Plan for Salinity and Water Quality (Tropical Rapid Appraisal of Riparian Condition (TRARC)). Queensland Government, Brisbane, Australia

36. Department of Sustainability, Environment, Water, Population and Communities (2012) Threat abatement plan to reduce the impacts on northern Australia's biodiversity by the five listed grasses. Australian Government, Canberra, Australia

37. Ward M (2014) Disease spread between domestic cattle and feral pigs: improving emergency preparedness. Meat & Livestock Australia, North Sydney, Australia

38. Iveson JB, Bradshaw SD, How RA, Smith DW (2014) Human migration is important in the international spread of exotic *Salmonella* serovars in animal and human populations. *Epidemiology & Infection* 142: 2281–2296

39. Sullivan K (2022) Minister rules out ban on Indonesian flights amid foot-and-mouth outbreak. *ABC News*, 14 July

40. Department of Climate Change, Energy, the Environment and Water (2012) Myrtle rust in natural ecosystems national workshop – summary of outcomes. Australian Government, Canberra, Australia

CHAPTER 12: CONSERVING THE BIODIVERSITY OF KIMBERLEY RAINFORESTS

1. Minister for Environment (2014) Biodiversity surveys assist Kimberley land management. Media release, 27 October. Government of Western Australia, Perth, Australia

2. McKenzie NL, Burbidge AA, Baynes A, Brereton RN, Dickman CR, Gordon G, Gibson LA, Menkhorst PW, Robinson AC, Williams MR, Woinarski JCZ (2007) Analysis of factors implicated in the recent decline of Australia's mammal fauna. *Journal of Biogeography* 34(4): 597–611

3. Burbidge AA, McKenzie NL, Brennan KEC, Woinarski JCZ, Dickman CR, Baynes A, Gordon G, Menkhorst PW, Robinson AC (2008) Conservation status and biogeography of Australia's terrestrial mammals. *Australian Journal of Zoology* 56: 411–422

4. Carwardine J, O'Connor T, Legge S, Mackey B, Possingham HP, Martin TG (2011) Priority threat management to protect Kimberley wildlife. CSIRO Ecosystem Sciences, Brisbane, Australia

5.   Kenneally KF, Edinger D, Coate K, Hyland B, How R, Schmitt L, Cowan M, Willing T, Done C (2003) The Last Great Wilderness - Exploring the Mitchell Plateau 2002. Landscope Expeditions Report no. 49. Western Australian Department of Conservation and Land Management, Perth, Australia

6.   Diamond JM (1976) Island biogeography and conservation: strategy and limitations. *Science* 193: 1027-1029

7.   Cowie RH, Holland BS (2006) Dispersal is fundamental to biogeography and the evolution of biodiversity on oceanic islands. *Journal of Biogeography* 33(2): 193-198

8.   MacArthur RH, Wilson EO (1967) *The theory of island biogeography*. Princeton University Press, New Haven, CT, USA

9.   Harris LD (1984) *The fragmented forest. Island biogeography theory and the preservation of biotic diversity*. University of Chicago Press, Chicago, IL, USA

10.   Department of Parks and Wildlife (2014) Kimberley science and conservation strategy: Key achievements of the landscape conservation initiative. Government of Western Australia, Perth, Australia

11.   Kerins S (2015) Kimberley conservation threatens to take a step back on Indigenous rights. *The Conversation*, 19 November

12.   McKenzie NL, Johnston RB, Kendrick PG (eds) (1991) *Kimberley rainforests of Australia*. Surrey Beatty & Sons, Chipping Norton, Australia

13.   Commonwealth of Australia (2011) Inclusion of a place in the National Heritage List: The West Kimberley. *Commonwealth of Australia Gazette*, 31 August

14.   Department of Sustainability, Environment, Water, Population and Communities (2012) West Kimberley national heritage place: a draft guide for landholders. Australian Government, Canberra, Australia

15.   Beard JS (1979) *Vegetation survey of Western Australia, sheet 1: Kimberley*. University of Western Australia Press, Nedlands, Australia

16.   Environmental Protection Authority (2001) EPA advice: Protection of tropical arid zone mangroves along the Pilbara coastline. Government of Western Australia, Perth, Australia

CHAPTER 13: ESTABLISHING A TROPICAL BIODIVERSITY RESEARCH CENTRE IN THE KIMBERLEY

1.   Ens EJ, Pert P, Clarke PA, Budden M, Clubb L, Doran B, Douras C, Gaikwad J, Gott B, Leonard S, Locke J, Packer J, Turpin G, Wason S (2015) Indigenous biocultural knowledge in ecosystem science and management: Review and insight from Australia. *Biological Conservation* 181: 133-149

2.   Pepper M, Keogh JS (2014) Biogeography of the Kimberley, Western Australia: a review of landscape evolution and biotic response in an ancient refugium. *Journal of Biogeography* 41(8): 1443-1455

3.   Laurie V (2010) *The Kimberley: Australia's last great wilderness*. UWA Publishing, Crawley, Australia

4.   Harris A (1983) Tropical centre needed. *The West Australian*, 1 October

5. Harris A (1983) Scientists want rain-forest centre. *The West Australian*, 19 December

6. Lowe P (2019) *The boab tree*. Backroom Press, Broome, Australia

7. Carmody J, Murphy H, Hill R, Catterall C, Goosem S, Dale A, Westcott S, Wellbergen J, Shoo L, Stoeckl N, Esparon M (2015) The importance of protecting and conserving the Wet Tropics: a synthesis of NERP Tropical Ecosystems Hub tropical rainforest research outputs 2011–2014. Report to the National Environmental Research Program. Reef and Rainforest Research Centre, Cairns, Australia

A Pheasant Coucal (*Centropus phasianinus*) drying off in monsoon rainforest, Broome. *Image: Tim Willing*

INDEX

a

C

e

f

g

h

i

j

k

n

t

Y

Z

Ngorlawuroo, also known as Mount Waterloo, is located in the Prince Regent National Park. It is a sandstone mesa, underlain with basalt, forming distinct benches. Both Mount Trafalgar (Ngayaanggananya) and Mount Waterloo were named by Lieutenant Philip Parker King in 1820 during his hydrographic surveys of the Australian coast. *Image: Mike Donaldson*